# HANDBOOK OF RESEARCH ON FOOD SCIENCE AND TECHNOLOGY

## Volume 2

Food Biotechnology and Microbiology

# HANDBOOK OF RESEARCH ON FOOD SCIENCE AND TECHNOLOGY

## Volume 2

Food Biotechnology and Microbiology

*Edited by*

**Mónica Lizeth Chávez-González, PhD**
**José Juan Buenrostro-Figueroa, PhD**
**Cristóbal N. Aguilar, PhD**

Apple Academic Press Inc.
3333 Mistwell Crescent
Oakville, ON L6L 0A2 Canada

Apple Academic Press Inc.
9 Spinnaker Way
Waretown, NJ 08758 USA

First issued in paperback 2021

*Exclusive worldwide distribution by CRC Press, a member of Taylor & Francis Group*
No claim to original U.S. Government works

**Handbook of Research on Food Science and Technology, Volume 2:
Food Biotechnology and Microbiology**
ISBN 13: 978-1-77463-126-3 (pbk)
ISBN 13: 978-1-77188-719-9 (hbk)
**Handbook of Research on Food Science and Technology, 3-volume set**
ISBN 13: 978-1-77188-721-2 (hbk)

### Library and Archives Canada Cataloguing in Publication

Handbook of research on food science and technology / edited by Mónica Lizeth Chávez-González, PhD, José Juan Buenrostro-Figueroa, PhD, Cristóbal N. Aguilar, PhD.

Includes bibliographical references and indexes.
Contents: Volume 2. Food biotechnology and microbiology.
Issued in print and electronic formats.
ISBN 978-1-77188-719-9 (v. 2 : hardcover).--ISBN 978-0-429-48783-5 (v. 2 : PDF)

1. Food industry and trade--Technological innovations. 2. Food--Research. 3. Food--Biotechnology. 4. Food--Composition. 5. Functional foods. I. Chávez-González, Mónica Lizeth, 1987-, editor II. Buenrostro-Figueroa, José Juan, 1985-, editor III. Aguilar, Cristóbal Noé, editor

| TP370.H36 2018 | 664'.024 | C2018-906012-3 | C2018-906013-1 |

CIP data on file with US Library of Congress

Apple Academic Press also publishes its books in a variety of electronic formats. Some content that appears in print may not be available in electronic format. For information about Apple Academic Press products, visit our website at **www.appleacademicpress.com** and the CRC Press website at **www.crcpress.com**

# CONTENTS

# ABOUT THE EDITORS

**Mónica Lizeth Chávez-González, PhD**
*Full Professor, School of Chemistry of the Universidad Autónoma de Coahuila, Mexico*

Mónica Lizeth Chávez-González, PhD, is a Full Professor at the School of Chemistry of the Universidad Autónoma de Coahuila, Mexico, where she develops her work in the Food Research Department. Dr. Chávez-González's experience is in the areas of fermentation processes, microbial biotransformation, enzyme production, valorization of food industrial wastes, extraction of bioactive compounds, and chemical characterization. She is a member of the Sociedad Mexicana de Biotecnología y Bioingeniería and the Asociación Mexicana para la Protección a los Alimentos affiliate of the International Association for Food Protection. She was awarded with the "Juan Antonio de la Fuente" medal for academic excellence and the "Ocelotl" prize for best tecnhological innovation purpose, both given by the Universidad Autónoma de Coahuila. She earned her PhD in Food Science and Technology with an emphasis on valorization of food industrial waste under the tutelage of Dr. Cristóbal N. Aguilar.

**José Juan Buenrostro-Figueroa, PhD**
*Researcher, Research Center for Food and Development, A.C., Mexico*

José Juan Buenrostro-Figueroa, PhD, is a Researcher at the Research Center for Food and Development, A.C., Mexico. Dr. Buenrostro has experience in bioprocess development, including microbial processes for enzyme production and recovery of bioactive compounds; valorization of agroindustrial byproducts, and extraction and characterization of bioactive compounds from plants and agroindustrial wastes. He has published 17 papers in indexed journals, five book chapters, four patent requests, and more than 45 contributions at scientific meetings. Dr. Buenrostro has been a member of S.N.I. (National System of Researchers, Mexico), the Mexican Society of Biotechnology and Bioengineering (SMBB), and the Mexican Society for Food Protection affiliate of the International Association for Food Protection.

He became a Food Engineer at the Antonio Narro Agrarian Autonomous University. He earned his MSc and PhD degrees in Food Science and Technology from the Autonomous University of Coahuila, Mexico, where he worked on the development of bioprocesses for the valorization of agroindustrial byproducts. He also worked at in the Biotechnology Department of the Metropolitan Autonomous University, Mexico City, Mexico.

### Cristóbal N. Aguilar, PhD
*Full Professor and Dean, School of Chemistry, Universidad Autónoma de Coahuila, Mexico*

Cristóbal N. Aguilar, PhD, is a Full Professor and Dean of the School of Chemistry at the Universidad Autónoma de Coahuila, Mexico. Dr. Aguilar has published more than 160 papers published in indexed journals, more than 40 articles in Mexican journals, as well as 16 book chapters, eight Mexican books, four international books, 34 proceedings, and more than 250 contributions in scientific meetings. Professor Aguilar is a member of the National System of Researchers of Mexico (SNI) and has received several prizes and awards, the most important are the National Prize of Research (2010) of the Mexican Academy of Sciences, the "Carlos Casas Campillo 2008" prize of the Mexican Society of Biotechnology and Bioengineering, the National Prize AgroBio-2005, and the Mexican Prize in Food Science and Technology from CONACYT-Coca Cola México in 2003. He is also a member of the Mexican Academy of Science, the International Bioprocessing Association (IFIBiop), and several other scientific societies and associations. Dr. Aguilar has developed more than 21 research projects, including six international exchange projects. He has been advisor of 18 PhD theses, 25 MSc theses, and 50 BSc theses.

He became a Chemist at the Universidad Autónoma de Coahuila, Mexico, and earned his MSc degree in Food Science and Biotechnology at the Autonomous University of Chihuahua, Mexico. His PhD degree in Fermentation Biotechnology was awarded by the Autonomous University of Metropolitana, Mexico. Dr. Aguilar also performed postdoctoral work at the Department of Biotechnology and Molecular Microbiology at Research Institute for Development (IRD) in Marseille, France.

# CONTRIBUTORS

**Cristóbal N. Aguilar**
Group of Bioprocesses and Bioproducts, Food Research Department, Chemistry School, Autonomous University of Coahuila, Blvd. V. Carranza e Ing. J. Cardenas V., Saltillo, Coahuila, CP 25280, Mexico, Tel.: 52 (844) 416-12-38, Fax: 52 (844) 415-12-38, +52 (844) 415-95-34, E-mail: cristobal.aguilar@uadec.edu.mx

**Miguel Á. Aguilar-González**
Center for Research and Advanced Studies of National Polytechnic Institute (CINVESTAV) Saltillo Unit, Av. Industrial Metalúrgica #1062, Parque Industrial Saltillo-Ramos Arizpe, Ramos Arizpe, Coahuila, México, C.P. 25900

**Antonio F. Aguilera-Carbó**
Animal Nutrition Department, Animal Science Division, Antonio Narro Agarian Autonomous University, CP 25315 Saltillo Coahuila, México, E-mail: miguelmem84@gmail.com

**Georgina Michelena Álvarez**
Cuban Institute for Research on Sugarcane Derivatives Cuba

**Juan A. Ascacio-Valdés**
Food Research Department, School of Chemistry, Autonomous University of Coahuila, Boulevard Venustiano Carranza and José Cárdenas s/n, República Oriente, Saltillo 25280, Coahuila, México

**Nagamani Balagurusamy**
Bioremediation Laboratory, Faculty of Biological Sciences, Autonomous University of Coahuila, Torreón, Coahuila, CP 27000, México

**Ruth E. Belmares**
Food Research Department, School of Chemistry, Autonomous University of Coahuila, Boulevard Venustiano Carranza and José Cárdenas s/n, República Oriente, Saltillo 25280, Coahuila, México

**Daniel Boone-Villa**
School of Medicine, North Unit UA de C, Calle de la Salud #714, Villa de Fuente, Piedras Negras, Coahuila, México, C.P. 26090.

**José Juan Buenrostro-Figueroa**
Research Center in Food and Development, A.C. Av. Cuarta Sur 3820, Fracc. Vencedores del Desierto, C.P. 33089, Cd. Delicias, Chihuahua, México, Tel.: +52 (555) 474-8400-Ext. 117, E-mail: jose.buenrostro@ciad.mx

**Sandra L. Castillo-Hernández**
School of Biology, Universidad Autónoma de Nuevo León, Monterrey Nuevo León, México

**Natarajan Chandrasekaran**
Centre for Nanobiotechnology, VIT University, Vellore–632014, Tamil Nadu, India, Tel.: +91-416-220262, E-mail: nchandrasekaran@vit.ac.in

**Monica L. Chávez-González**
Group of Bioprocesses and Bioproducts, Food Research Department, Chemistry School,
Autonomous University of Coahuila, Blvd. V. Carranza e Ing. J. Cardenas V., Saltillo, Coahuila,
CP 25280, Mexico

**Reynaldo De la Cruz**
School of Engineering & Sciences, Monterrey Institute of Technology, Ave.
Eugenio Garza-Sada 2501, CP 64849, Monterrey, Nuevo León, México

**Marisol Cruz-Requena**
Research Center, Identification, Guard and Microbiological Analysis, , Rio de Janeiro 590,
Colonia Latinoamericana, CP 25270, Saltillo Coahuila, México

**Ileana Mayela Moreno Dávila**
Group of Bioprocesses and Bioproducts, Food Research Department, Chemistry School,
Autonomous University of Coahuila, Blvd. V. Carranza e Ing. J. Cardenas V., Saltillo, Coahuila,
CP 25280, Mexico

**René Díaz-Herrera**
Group of Bioprocesses, Food Research Department School of Chemistry,
Universidad Autónoma de Coahuila, Saltillo, Coahuila, CP 25280, México

**Sandra C. Esparza-González**
School of Medicine, Universidad Autónoma de Coahuila, Saltillo, Coahuila, México

**Rosario Estrada-Mendoza**
Food Research Department, School of Chemistry, Universidad Autónoma de Coahuila,
Boulevard Venustiano Carranza and José Cárdenas s/n, República Oriente, Saltillo 25280,
Coahuila, México

**Adriana C. Flores-Gallegos**
Food Research Department, School of Chemistry, Universidad Autónoma de Coahuila,
Boulevard Venustiano Carranza and José Cárdenas s/n, República Oriente, Saltillo 25280,
Coahuila, México

**José Daniel García**
Food Research Department, Autonomous University of Coahuila, Mexico

**Alfredo I. García-Galindo**
Group of Bioprocesses and Bioproducts, Food Research Department, Chemistry School,
Autonomous University of Coahuila, Blvd. V. Carranza e Ing. J. Cardenas V., Saltillo, Coahuila,
CP 25280, Mexico

**Eduardo García-Martínez**
Animal Nutrition Department, Animal Science Division, Antonio Narro Agarian Autonomous
University, CP 25315 Saltillo Coahuila, México, E-mail: miguelmem84@gmail.com

**Heliodoro de la Garza-Toledo**
Antonio Narro Agrarian Autonomous University. Blvd. Antonio Narro 1923 Col. Buenavista,
Saltillo, Coahuila, México, C.P. 25315

**Ricardo Gómez-García**
Group of Bioprocesses and Bioproducts, Food Research Department, Chemistry School,
Autonomous University of Coahuila, Blvd. V. Carranza e Ing. J. Cardenas V., Saltillo, Coahuila,
CP 25280, Mexico

**José Luis Martínez Hernández**
Food Research Department, Autonomous University of Coahuila, Mexico,
E-mail: jose-martinez@uadec.edu.mx

**María G. Hernández-Ángel**
Food Science and Technology Deparment, Autonomous University of Coahuila, Chemistry School,
Blvd. Venustiano Carranza y José Cárdenas Valdez S/N Col. Republica Oriente, Saltillo Coahuila,
C.P. 25280

**Anna Iliná**
Food Research Department, Autonomous University of Coahuila, Mexico

**Cristian Martínez-Ávila**
School of Agronomy, Autonomous University of Nuevo Leon, Francisco Villa S/N,
Col. Ex-Hacienda el Canadá, Gral. Escobedo, Nuevo León, México, C.P. 66050

**Miguel A. Medina-Morales**
Animal Nutrition Department, Animal Science Division, Antonio Narro Agarian Autonomous
University, CP 25315 Saltillo Coahuila, MéxicoCP 25315 Saltillo Coahuila, México,
E-mail: miguelmem84@gmail.com

**Miguel Mellado**
Animal Nutrition Department, Animal Science Division, Antonio Narro Agarian Autonomous
University, CP 25315 Saltillo Coahuila, México, CP 25315 Saltillo Coahuila, México,
E-mail: miguelmem84@gmail.com

**Amitava Mukherjee**
Centre for Nanobiotechnology, VIT University, Vellore–632014, Tamil Nadu, India

**Erika Nava-Reyna**
National Institute for Forestry, Agriculture and Livestock Research (INIFAP) CENID-RASPA,
Mexico

**Emilio Ochoa-Reyes**
Research Center in Food and Development, A.C. 31570, Cd. Cuauhtémoc, Chihuahua, Mexico

**Arely Prado-Barragán**
Biotechnology Department, Autonomous Metropolitan University, 09340, Iztapalapa,
Mexico City, México

**Raul Rodríguez-Herrera**
Food Research Department, School of Chemistry, Universidad Autónoma de Coahuila,
Boulevard Venustiano Carranza and José Cárdenas s/n, República Oriente, Saltillo 25280,
Coahuila, México, E-mail: raul.rodriguez@uadec.edu.mx

**Romeo Rojas**
School of Agronomy, Research Center and Development for Food Industries,
Autonomous University of Nuevo Leon, General Escobedo, Nuevo León, México

**Rosa Salas-Valdez**
Group of Bioprocesses and Bioproducts, Food Research Department, Chemistry School,
Autonomous University of Coahuila, Blvd. V. Carranza e Ing. J. Cardenas V., Saltillo,
Coahuila, CP 25280, Mexico

**Leonardo Sepúlveda**
Food Research Department, Autonomous University of Coahuila, Saltillo, Coahuila, México

**J. S. Swathy**
Centre for Nanobiotechnology, VIT University, Vellore–632014, Tamil Nadu, India

**Julio C. Tafolla-Arellano**
Research Center in Food and Development, A.C. Vegetal Origin Food Technology Coordination, 83304, Hermosillo, Sonora, México

**Juan M. Tirado-Gallegos**
Research Center in Food and Development, A.C. 31570, Cd. Cuauhtémoc, Chihuahua, Mexico

**Janeth Margarita Ventura-Sobrevilla**
School of Health Sciences UA de C, Calle de la Salud #714, Villa de Fuente, Piedras Negras, Coahuila, México. C.P. 26090, E-mail: janethventura@uadec.edu.mx

# ABBREVIATIONS

| | |
|---|---|
| anti-HSV | antiviral against the herpes simplex virus |
| AOX | alcohol oxidase |
| ATP | adenosine triphosphate |
| BHA | butylated hydroxyanisole |
| BHT | butylated hydroxytoluene |
| BPDE | benzo [a] pyrene–7,8-diol–9,10-epoxide |
| CBD | Convention on Biological Diversity |
| CO | carbon monoxide |
| $CO_2$ | carbon dioxide |
| DLS | dynamic light scattering |
| DOE | design of experiments |
| DSSC | dye-sensitized solar cells |
| EAE | enzyme-assisted extraction |
| EAEC | *E. coli* enteroagregative |
| ECG | epicatechin-3-gallate |
| EGCG | epigallocatechin-3-gallate |
| EIP | emulsion inversion point |
| EPEC | enteropathogenic *E. coli* |
| ETEC | enterotoxigenic *E. coli* |
| EU | European Union |
| FAE | fermented-assisted extraction |
| FBD | Food-borne diseases |
| FDA | Food and Drug Administration |
| GAP | glyceraldehyde dehydrogenase |
| GMC | genetically modified corns |
| GMO | genetically modified organism |
| HC | hemorrhagic |
| HIV | human immunodeficiency virus |
| HLB | hydrophile- lipophile balance |

| LBGMO | Law of Biosecurity on Genetically Modified Organisms |
|---|---|
| LSW | lifshitz- slezov and wagner |
| MAE | microwave-assisted extraction |
| MGM | genetically modified maize |
| NDGA | nordihydroguaiaretic acid |
| NIH | National Institutes of Health |
| PCS | photon correlation spectroscopy |
| PG | propyl gallate |
| RSM | response surface methodology |
| SEDDS | self-emulsifying drug delivery systems |
| SEM | scanning electron microscopy |
| SLS | static light scattering |
| SNEDDS | self-nano emulsifying drug delivery systems |
| SPC | single-cell protein |
| SSC | solid-state culture |
| SSF | solid-state fermentation |
| TBHQ | tert-butyl hydroquinone |
| TEM | transmission electron microscopy |
| US | United States |
| WHO | World Health Organization |

# PREFACE

Human population is growing dramatically, and it will probably reach 10 billion world inhabitants sooner than estimated. It implies important challenges never faced before by humans who need to organize and work on the development of mega-efforts to ensure universal access to health care, food, water, sanitization, energy, education, and housing. These challenges, natural or man-made, obligate the scientific community to proactively seek new breakthrough food and nutrition solutions to ensure global food sustainability and nutrition security in the future. To achieve this, innovative solutions need to be considered throughout the whole food chain, inclusive of food choices and dietary patterns, in order to make significant improvements in the food supply, nutritional, and health status. In the case of foods, innovations in food processing techniques can significantly contribute to meeting the needs of the future world population with respect to quality, quantity, and sustainability of food intake.

Those in academe and industry focused on food science and technology are constantly redefining their traditional forms for new ways to face the threats of the twenty-first century, which is marked by multiple unprecedented environmental challenges that could threaten human survival. The combined impact of climate change, energy and water shortages, environment pollutants, shifting global population demographics, food safety, and growing disease pandemics all place undue stress on the planet's food system, already in a sensitive balance with its ecosystem.

Any changes to the food supply inevitably impact food, nutrition, and health trends and policies, particularly pertaining to food production, agricultural practices, dietary patterns, nutrition, and health guidance and management. As a result, there is an urgent need to find alternative solutions to improve the efficiency and sustainability in the food supply chain by reducing food waste and enhancing nutritional qualities of foods through the addition of nutraceuticals to prepare functional foods and intelligent foods.

The Food Research Group of the School of Chemistry at Universidad Autonoma de Coahuila (DIA-UAdeC) celebrates 25 years of existence and hard work, a period in which it has undergone a tremendous transformation in order to provide solutions and new technological alternatives to the problems demanded by the region, the country, and some international elements. To achieve this, the group grew in the number of researchers, and therefore various lines of research are studied. Today the DIA-UAdeC is formed by research groups in Bioprocesses and Bioproducts, Biorefineries, Biocontrol, Natural Products, Molecular Biology and Ommic Sciences, Glic-Biotechnology, Nano-Bioscience, Edible Coatings, Films and Membrane Technology, Food Engineering, Emerging Processing Technology, Food Science, and Functional Foods.

Two important Mexican postgraduate programs in Food Science and Technology are offered to Mexican and foreign students to whom the National Council of Science and Technology of Mexico (CONACYT) offers scholarships to carry out their MSc or PhD programs.

The consolidation of national scientific cooperation has allowed the prolongation and substantially improvement of the generation and application of knowledge with the Autonomous Metropolitan University, the Autonomous University of Chihuahua, the Autonomous University of Tamaulipas, the Autonomous Agrarian University "Antonio Narro," the Autonomous University of Nuevo León, the University of Colima, the Technological Institute of Durango, the Technological Institute of Ciudad Valles, the Technological Institute of Monterrey, and the centers of research CIQA, CINVESTAV, CIMAV, CIAD, CICATA, CIATEJ, among others.

Strong linkages of international cooperation with institutions and research centers around the world have been established and are now generating important results in the framework of scientific and technological cooperative projects and programs. They highlight their research partnerships with the University of Minho (Portugal); the University of Vigo (Spain); the University of Georgia (USA); the University of Marseille (France); the University of Valle and the National University of Colombia; the National University Nacional de Rosario, the National University of Rio Cuarto and the National University of La Plata (Argentina); Kannur University (India); Federal University of Pernambuco (Brazil); University of Torino (Italy); Jacobs University (Germany); Gachon University (Korea); and other important world-quality research centers including INL (Portugal); IRD and IMBE (France); ICIDCA (Cuba), and the Jawaharlal Nehru Tropical Botanic Garden & Research Institute (India).

For this reason, the research group has organized itself to celebrate its 25th anniversary by publishing a book that reflects the scientific and technological contributions in the field of food Science and technology generated by scientists of the DIA-UAdeC and some of its collaborators.

This *Handbook of Research in Food Science and Technology* consists of three volumes; (i) Food Technology and Chemistry, (ii) Food Biotechnology and Microbiology, and (iii) Functional Foods and Nutraceuticals, all of which will highlight the current trends and knowledge regarding the most recent innovations, emerging technologies, and strategies based on food design on a sustainable level. The handbook includes relevant information on the modernization of food industries, emerging technologies, sustainable packaging, food bioprocesses, food fermentation, food microbiology, functional foods, nutraceuticals, natural products, nano- and micro-technology, healthy product composition, innovative processes/bioprocesses for utilization of byproducts, development of novel preservation alternatives, extending the shelf life of fresh products, and alternative processes requiring less energy or water, among other topics.

# CHAPTER 1

# VALORIZATION OF AGROINDUSTRIAL BYPRODUCTS: BIOTECHNOLOGICAL PROCESSES FOR EXTRACTION OF PHENOLIC COMPOUNDS

ARELY PRADO-BARRAGÁN,[1] ROMEO ROJAS,[2]
EMILIO OCHOA-REYES,[3] JUAN M. TIRADO-GALLEGOS,[3]
JULIO C. TAFOLLA-ARELLANO,[4] and
JOSÉ JUAN BUENROSTRO-FIGUEROA[5]

[1] Biotechnology Department, Metropolitan Autonomous University, 09340, Iztapalapa, Mexico City, México

[2] School of Agronomy, Research Center and Development for Food Industries, Autonomous University of Nuevo Leon, General Escobedo, Nuevo León, México

[3] Research Center in Food and Development, A.C. 31570, Cuauhtémoc, Chihuahua, Mexico

[4] Research Center in Food and Development, A.C. Vegetal Origin Food Technology Coordination, 83304, Hermosillo, Sonora, México

[5] Research Center in Food and Development, A.C. 33089, Delicias, Chihuahua, México, Tel.: +52 (555) 474-8400-Ext. 117, E-mail: jose.buenrostro@ciad.mx

## ABSTRACT

During the processing of raw vegetables and fruit, substantial quantities of agroindustrial byproducts, such as pods, peel, pulp, stones, and seeds, are generated. These materials are commonly disposed of in local landfills, due to the lack of appropriate infrastructure and any established commercial use. Moreover, agroindustrial byproducts possess a high concentration of biodegradable organic matter, which promote leachate and methane emissions, resulting in pollution. Valorization is a recent concept in the field of industrial residues management, with increasing practice. The agroindustrial byproducts valorization aims to recover fine chemicals and obtain other metabolites, through chemical and biotechnological processes. The composition of the agroindustrial byproducts represents an excellent source of high added-value ingredients that can be recovered and valorized in the food, chemical, and pharmaceutical industries. The recovery of phenolic compounds from byproducts is of particular interest, due to their ability to promote benefits for human and animal health. Although various extraction methods for phenolic compounds have been reported, biotechnological processes have received considerable attention by several research groups. Currently, the enzyme- and fermentation-assisted extraction methods are intensively studied. In this chapter, an overview of the trends in the extraction of phenolic compounds by biotechnological procedures and their economic feasibility is presented.

## 1.1  INTRODUCTION

Agroindustrial byproducts represent serious environmental pollution and worldwide economic concerns, due to improper handling [1]. Some approaches to overcome this concern include the utilization of crop residues and agroindustrial byproducts, to develop systems of livestock feeding and composting [2, 3]. However, these methods are limited to particular agroindustrial byproducts and insufficient to resolve the issue. Thus, the valorization of agroindustrial byproducts has become an alternative approach to minimize the environmental problem and alleviate economic concerns because the agro-residues are rich in bioactive and nutritional compounds with human benefits, as well as environmental and industrial profits. Some of these biological properties reported are anti-proliferative, antidiabetic, anticancer, antioxidant, antimicrobial, anti-inflammatory, and antiviral activities. The valorization of agroindustrial byproducts demands biotechnological processes for efficient and maximal extraction of the bioactive compounds. In this regard, several non-conventional methods extraction techniques, have been developed to overcome the limitations of traditional approaches. For instance, the conventional Soxhlet extraction is usually time-consuming and results in a low extraction efficiency. In comparison, the non-conventional extraction methods like enzyme-assisted extraction (EAE) and fermented-assisted extraction (FAE) are more environmentally friendly, reduce the time and solvent consumption, and increase the extraction yield and quality [4, 5].

## 1.2 AGROINDUSTRIAL BYPRODUCTS

Agroindustry refers to the combined use of agricultural and industrial processes or methods for the transformation of raw agricultural material or byproducts into value-added products like food, chemicals, and fertilizers. An agroindustry is an enterprise, where plant materials are transformed and preserved using physical, chemical, mechanical, and nowadays, biological methods. The nature of the process or degree of transformation depends on the final use of the raw material, and it can be categorized by the transformation level (Table 1.1).

The agroindustrial waste or byproducts are generated during the transformation process of agricultural products and include straws, stems, stalks, leaves, husks, shells, peels, seeds, pulps, bagasse, and other [6].

### 1.2.1 CHEMICAL COMPOSITION

The agroindustry byproducts contain cellulose, hemicellulose, lignin, sugars, and carbon and nitrogen sources (Tables 1.2–1.4). The valorization

**TABLE 1.1**  Agroindustrial Classification by Transformation Level of Raw Materials

| Level | Transformation | Examples |
|-------|----------------|----------|
| I | Preparation: Cleaning, Grading, storage | Fresh fruits, fresh vegetables, eggs. |
| II | Ginning, Milling, Cutting, Mixing | Cereals, spices, animal feeds, cotton, jute, flour, lumber. |
| III | Cooking, Pasteurization Canning, Dehydration, Freezing, Weaving, Extraction, | Canned products: fruits, vegetables, sugar, dairy products, vegetable oils, beverages. |
| IV | Chemical treatments, Texturization | Textured vegetables, instant meals. |

of organic byproducts is a valuable approach to harness the nutritious composition of agricultural waste, better manage environmental pollution and attain income for the agro community. The valuable properties of agricultural byproducts are of great interest as materials to produce several, safe, added-value, and environmentally friendly products. Agricultural byproducts are highly available biomass that can be used as alternative raw materials for the food, pharmacy, and detergent industries, as well as for biocomposite and biomedical components and others.

## 1.2.2   GENERATION

The industrial processing of the edible agricultural products generates large volumes of byproducts, causing a severe disposal problem [9]. In Europe, about $2.5 \times 10^8$ ton per year of agroindustrial byproducts are generated [10]. These derivates include the bagasse and peels produced from the beverages and juice industries, coffee pulp from the coffee industry, and various kinds of husks from the cereal industry. Most of these residues have a high nutritional and chemical composition, attracting interest from academic and industrial researchers as potential sources of bioactives, platform chemicals and other value-added products. Due to a rich carbohydrate composition, these residues are easily assimilated by microorganisms and, consequently, could be appropriate for use as raw materials in the production of industrially relevant compounds under fermentation processes [9].

Table 1.5 shows the quantity and quality of some byproducts that are highly valuable regarding their nutritional composition and amount produced. In developing countries, agroindustrial byproducts like olive cake and brewer's grain, as well as high moisture agroindustrial byproducts (e.g., citrus pulp, sugar beet pulp, tomato pulp), are of high nutritional value and can be utilized as straight feeds and/or supplements for upgrading the nutritional value and as raw materials to produce value-added components.

## 1.2.3   PROBLEMS GENERATED BY FINAL DEPOSITION OF AGROINDUSTRIAL BYPRODUCTS

Deposition of the agricultural byproducts is a critical problem that needs to be solved for the safety of the global environment. Dumping or burning wastes or agroindustrial byproducts present potential air and water pollution

**TABLE 1.2** Chemical Composition of Agro-Industrial Byproducts [7]

| Agroindustrial byproducts | Moisture | Total solids | Chemical composition (% w/w dry weight) | | | | | | | |
|---|---|---|---|---|---|---|---|---|---|---|
| | | | Ash | Cellulose | Hemicellulose | Lignin | Total sugars | Total carbon | Total nitrogen |
| Corn stalks | 1.92 | 97.78 | 10.8 | 61.2 | 19.3 | 6.9 | 0.22 | 50.3 | 1.05 |
| Rice straw | 1.83 | 98.62 | 12.4 | 39.2 | 23.5 | 36.1 | 0.071 | 41.8 | 0.457 |
| Sawdust | 1.12 | 98.54 | 1.2 | 45.1 | 28.1 | 24.2 | 0.025 | 37.8 | 0.24 |
| Sugarcane bagasse | 8.34 | 91.66 | 1.9 | 30.2 | 56.7 | 13.4 | 0.55 | 36.45 | 0.448 |
| Sugar beet waste | 12.4 | 87.5 | 4.8 | 26.3 | 18.5 | 2.5 | 0.83 | 44.5 | 1.84 |

**TABLE 1.3**   Proximate Chemical Composition of Fruit Peels (% of Dry Peel) [8]

| Fruit | Yield (g/100 g of fresh of fruit) | Crude protein | Lipids | Ash | Crude fiber | Carbohydrates |
|-------|-------|-------|-------|-------|-------|-------|
| Apple | 10.20 | 2.80 | 9.96 | 1.39 | 13.95 | 59.96 |
| Banana | 33.81 | 10.44 | 8.40 | 12.45 | 11.81 | 43.40 |
| Mango | 9.94 | 5.00 | 4.72 | 3.24 | 15.43 | 63.80 |
| Orange | 14.27 | 9.73 | 8.70 | 5.17 | 14.19 | 53.27 |
| Pawpaw | 10.21 | 18.06 | 5.47 | 10.22 | 12.16 | 37.49 |
| Pineapple | 9.17 | 5.11 | 5.31 | 4.39 | 14.80 | 55.52 |
| Pomegranate | 11.69 | 3.46 | 3.36 | 6.07 | 17.63 | 59.98 |
| Watermelon | 6.44 | 12.42 | 12.61 | 5.03 | 26.31 | 32.16 |

**TABLE 1.4**   Mineral Composition of Fruit Peels [8]

| Fruit | Elements (mg/100 g dry peel) | | | |
|-------|-------|-------|-------|-------|
|  | Calcium | Zinc | Iron | Manganese |
| Apple | 14.895 | 0.95 | 25.63 | 1.28 |
| Banana | 19.86 | 1.72 | 15.15 | 9.05 |
| Mango | 60.63 | 0.66 | 12.79 | 4.77 |
| Orange | 162.03 | 6.84 | 19.95 | 1.34 |
| Pawpaw | 11.44 | 2.68 | 27.61 | 0.52 |
| Pineapple | 8.30 | 6.46 | 25.52 | 5.32 |
| Pomegranate | 52.92 | 0.98 | 9.22 | 0.58 |
| Watermelon | 11.21 | 3.78 | 45.58 | 1.25 |

issues. It is estimated that burning biomass, such as wood, leaves, trees, and grasses—including agricultural waste—produces 40% carbon dioxide ($CO_2$), 32% carbon monoxide (CO), 20% particulate matter, and 50% polycyclic aromatic hydrocarbons released into the environment around the globe [12]. Although agricultural waste burning is not an environmentally acceptable form of agricultural management, it is a frequent practice and a public health concern for many reasons.

The main techniques to dispose of agricultural byproducts are landfill and incineration. However, inappropriate management of landfill will result in emissions of methane and $CO_2$ [13], while incineration involves the subsequent formation and releases of pollutants and secondary wastes

**TABLE 1.5** Nutrients Composition From Agroindustrial Byproducts [11]

| | Amount (ton/yr) | ME (Mcal/kg) | Mcal (106 ton/yr) | CP,% | CP (ton/yr) | CF,% | CF (ton/yr) |
|---|---|---|---|---|---|---|---|
| Bread/cake/dried bakery waste | 482 | 3.23 | 1.56 | 9.80 | 47.24 | 1.20 | 5.78 |
| Brewer's grain | 640 | 2.46 | 1.57 | 27.10 | 173.44 | 13.20 | 84.48 |
| Citrus pulp dried citrus meal | 2,545 | 2.83 | 7.20 | 6.50 | 165.42 | 13.10 | 333.40 |
| Corn residues husks and leaves | 360 | 2.30 | 0.83 | 7.30 | 26.28 | 30.20 | 108.72 |
| Grapes, skin, and seeds | 70 | 0.93 | 0.06 | 11.80 | 8.26 | 29.00 | 20.30 |
| Olive Cake[1] skin, pulp, seed after oil extract | 75 | 3.55 | 26.62 | 5.00 | 3.75 | 15.00 | 11.25 |
| Poultry Wastes with Litter | 980 | 2.22 | 2.17 | 21.90 | 214.62 | 14.40 | 141.12 |
| Sesame Hull | 1,109 | 2.76 | 3.06 | 45.50 | 504.60 | 5.70 | 63.21 |
| Sugar Beet Molasses | 12,000 | 1.44 | 17.28 | 6.60 | 792.00 | 0.00 | 0.00 |
| Sugar Beet Pulp | 13,500 | 2.58 | 34.08 | 8.80 | 1188.00 | 18.00 | 2430.00 |
| Sun Flower Residues | 6,000 | 2.10 | 12.60 | 6.70 | 402.00 | 35.10 | 2106.00 |
| Wheat Bran | 32,200 | 2.37 | 76.30 | 15.20 | 4894.40 | 10.00 | 3220.00 |
| Whey dried product | 78 | 2.87 | 0.22 | 16.70 | 13.00 | 0.20 | 0.16 |
| Yeast Brewers | 117 | 2.87 | 0.33 | 43.80 | 51.25 | 2.90 | 3.40 |
| Total | 70,156 | | 183.90 | | 8484.26 | | 8527.82 |

ME: Metabolizable energy; CP: Crude Protein; CF: Crude fiber.

[1]Data from: Feeding Ensiled Crude Olive Oil Cake, M. Hadjipanayiotou, Livestock Production Science 59 (1999) 61–66; All other Data from: US Canadian Tables of Feed Composition, NRC ME in MCal/kg for Ruminants; Crude Protein, and Crude Fiber Data is expressed 100% D.

(e.g., dioxins, furans, acid gases), in addition to numerous contaminating particles [14], which cause serious environmental and health risks. For these reasons, there is an urgent need to seek a feasible strategy to discard, by concomitantly, reusing, and harnessing the nutritional, bioactive, and chemical value of agricultural byproducts. Moreover, inexpensive and readily available procedures to use agrifood industry waste is highly cost-effective and minimizes environmental impact. One of the most valuable approaches is to recover the bioactive constituents, particularly, enzymes and phenolic compounds, capitalizing on their potential in the food, pharmaceutical [15], as well as cosmetics industries [16]. Thus, the use of the agricultural byproducts as sources of bioactive compounds may be of considerable economic benefit and has become increasingly attractive.

## 1.3  VALORIZATION: BIOTECHNOLOGICAL PROCESSES

Agroindustrial byproducts represent a serious environmental problem and the major disposal issue for the industry concerned, causing expenses for their proper disposal and an increase in pollution, due to the high content of organic substances, which might represent legal problems. However, these wastes are a potential source of bioactive compounds, including small peptides, oils, poly/oligosaccharides, phenolic compounds and other useful ingredients [17]. Phenolic compounds are widely known for their biological properties, attributed to their antiproliferative, antidiabetic, anticancer, antioxidant, antimicrobial, anti-inflammatory, and antiviral properties, among others [18]. Recently, biotechnological processes have gained importance for the production or extraction of phenolic compounds from natural sources, providing economic and environmental advantages, as well as potential applications in the food, chemical, and pharmaceutical industries [1]. Previous studies have focused on releasing the phenolic compounds either by EAE, an enzymatic pretreatment (before extraction) to degrade the plant cell walls, or FAE, whereby a fermentation process (e.g., submerged or solid-state), allows obtaining the phenolic compound derivatives by fungal hydrolysis of macromolecules into smaller molecules.

### 1.3.1   EAE

The EAE of phenolic compounds from plants and agroindustrial byproducts has been widely studied. This process continues to gain attention due to the need for eco-friendly extraction technologies. Diverse phytochemical compounds are dispersed in the plant cell cytoplasm and intricately bound within the polysaccharide-lignin matrix by hydrogen or hydrophobic bonds, which are not released by conventional solvent extraction [17].

The EAE implies the use of enzymes to hydrolyze and degrade plant cell wall constituents, to improve the release and recovery of intracellular components, such as proteins, oils, and phenolic compounds [19]. The studies about EAE for phenolic compounds extraction have reported the use of cellulases, pectinases, amylases, glucosidases, xylanases, and other enzymes (Table 1.6). An important factor in the extraction process is the solubility of the target compound [20]. Low solubility leads to low extraction yield and requires large amounts of solvents, increasing the cost and efficiency. The EAE process improves the solubility of phenolic compounds, enhancing its extraction yield. Sahne et al. [21] reported the use of α-amylase and glucosidase to degrade the cell walls of turmeric (*Curcuma longa* L.), releasing curcumin, a bioactive compound with a broad spectrum of pharmacological effects, but a poor aqueous solubility that limits its medical applications. Enzymatic treatment improved the solubility of curcumin, leading to an increase in the ionic liquid extraction yield from 3.58% (without EAE) to 5.73% (with EAE). Therefore, in this instance, EAE might reduce or even exclude the use of a solvent for extraction, because water is used as the extraction solvent, which has many advantages compared to the use of hazardous solvents in conventional extraction.

## 1.3.1.1 ADVANTAGES AND DISADVANTAGES

EAE is a promising process for facilitating the release and recovery of a broad range of bioactive compounds. Several advantages and disadvantages have been reported, as listed below [19, 43].

**Advantages:**

- Enzymes are widely used on an industrial scale;
- Low energy consumption;
- High-extraction selectivity, due to enzyme specificity;

TABLE 1.6  Use of Enzymes for Extraction of Phenolic Compounds From Agroindustrial Byproducts

| Source | Enzyme | Compound | Highlights | Ref. |
|---|---|---|---|---|
| *Forsythia suspense* seeds | Cellulase Pectinase Protease | Seed oil and phenolic acids | Oil had quite high contents of unsaturated fatty acid (92.49%) and total phenolics (909.65 mg gallic acid/kg oil). | [22] |
| Thyme leaves Rosemary leaves | Cellulase Hemicellulase Mixture | Carvacrol Cineole | Enhancing yield and antimicrobial activity of the essential oil, as well as the content of carvacrol and 1,8-cineole. | [23] |
| Turmeric powder | α-amylase Glucoamilase Mixture | Oleoresin Curcumin | Maximal extraction yield was observed with glucoamylase treatment (22.5% and 31.83% for oleoresin and curcumin yield, respectively). | [24] |
| *Polygonum cuspidatum* roots | Pectinex® Viscozyme® | Resveratrol | Yield was significantly increased using pectinex. Resveratrol yield of 11.88 mg/g was obtained under optimized conditions. | [25] |
| *Eucommi ulmoides* leaves | Cellulase Pectinase Dextranase | Chlorogenic acid | Cellulase facilitated the cell wall degradation and improve the permeability of ionic liquids solution, promotes a best extraction. | [26] |
| Blackcurrant pomace | Viscozyme® L CelluStar® XL | Fatty acids Phenolic compounds | Yield of 2.73-fold higher using enzyme treatment than that found with other extraction methods. Extracts rich in poli-unsaturated fatty acids (linoleic 46.89% and linolenic 14.02%). | [27] |
| Buckwheat hulls | Viscozyme® L | Phenolic compounds | Extraction yield was increased 4–5 times, showing better antioxidant activity. | [28] |
| Bay leaves | Cellulase Hemicellulase Xylanase Mixture of them | Essential oil and phenolic compounds | Enhance of 243, 227 and 240% in essential oil yield in samples treated with cellulase, hemicellulose, and xylanase, respectively. Release of phenolic compounds and antioxidant activity were increased. | [29] |
| Winemaking byproducts | Pronase Viscozyme | Phenolic compounds | Increase in amount of soluble phenolics content, while decreasing the insoluble-bound phenolics. Increased in antioxidant activity by DPPH and ABTS radical scavenger. | [30] |

**TABLE 1.6** *(Continued)*

| Source | Enzyme | Compound | Highlights | Ref. |
|---|---|---|---|---|
| Ginger root | α-amylase Viscozyme | Gingerol | Yield was improved 1.33 and 1.9 times to oleoresin and gingerol, respectively. | [31] |
| Grape pomace | Pectinase and cellulase | Anthocyanin and other phenolic compounds | Pre-treatment with hot water prior to enzyme treatment increased the yield significantly. | [32] |
| Citrus peels (5 varieties) | Cellulase® MX Celullase® CL Kleerase® AFP | Total phenolic contents | The use of cell wall-degrading enzymes increases the extraction yield at 65.5% | [33] |
| Apple peel | Cellulase | Phenolics | Phenolic release was increased. | [34] |
| Kinnow peel | α-L-Rhamnosidase | Naringin | Enzyme catalyzes the cleavage of terminal rhamnosyl groups from naringin to yield prunin and rhamnose. | [35] |
| Grape pomace | Mixture of pectinase and cellulase | Phenolic acids, flavonoids, and anthocyanins | Enzymatic treatment improved the extraction yields at 91.9%, 92.4% and 64.6% for phenolic acids, flavonoids, and anthocyanins, respectively. | [36] |
| Grape pomace | Celluclast® 1.5 L Pectinex® Ultra Novoferm® | Phenolic acids | Increment of antioxidant activity (86.8, 82.9 and 90% with celluclast, pectinex, and novoferm) associated with the release of *O*-coumaric acid. | [1] |
| *Gingko biloba* leaves | Cellulase | Flavonoids | Enzyme treatment could transglycosilate flavonol aglycones into more polar glucosides, increasing its solubility, which improves the extraction up to 31%. | [20] |
| Bilberry skin | Pectinex® Panzym® Pro Panzym® BE | Anthocyanidins (Delphinin and cyanidin glycosides) | Phenolic compounds content was increased up to 4.7 times. Also, EAE decreased and inhibit the microbial growth. | [37] |
| Black pepper | α-amylase | Piperine | Increase in yield and phytochemical properties of extracts. | [38] |

**TABLE 1.6** *(Continued)*

| Source | Enzyme | Compound | Highlights | Ref. |
|---|---|---|---|---|
| Tomato waste | Pectinase Cellulase | Lycopene | Total carotenoid and lycopene extraction yield were 6-fold and 10-fold increase. | [39] |
| Pomegranate peels | Cellulase Pectinex Viscozyme Kemzyme Alcalase Cocktail | Phenolic acids | Extraction yield was increased 2-fold with a high antioxidant activity. | [40] |
| *Ulmus pumila* barks | Cellulase Pectinase β-glucosidase | Phenolic acids | Higher productivity of total phenolic and antioxidant activity. | [41] |
| Tomato peel | Peclyve PR Cellulyve 50 LC | Lycopene | Lycopene recovery was increased using mixed enzyme preparations, reaching up to 18-fold in extraction yields. | [42] |

- Reduced consumption of harmful chemicals like extraction solvents, particularly hexane, which render the product highly unsuitable for human consumption;
- No thermal decomposition of thermolabile compounds;
- Reduced extraction time;
- Efficient: enhances the yield and quality;
- Compared to solvent extraction, EAE is much safer, environmental-friendly, and economical;
- Low cost.

**Disadvantages:**
- May require multiple steps, using multi-enzyme preparations;
- Cost: mostly high priced commercial grade enzymes have been used;
- Requires the control of several process variables (temperature, pH, agitation, substrate concentration, solid/water ratio) that increase the cost of the process.

Additionally, EAE is an efficient and eco-friendly green process that can be used for the optimal extraction and recovery of high-value phenolic and other antioxidant compounds with natural chemo-preventive and nutraceutical potential, from any agroindustrial by-product [40].

## 1.3.1.2   PROCESS PARAMETERS

The enzyme type and concentration, the particle size of the material, solid/water ratio, pH, temperature, and process time are among several factors that must be considered, to identify the best extraction conditions [17, 24]. Optimal EAE conditions allow a several-fold increase in the extraction yield (Table 1.1). Due to the particular properties and phenolic composition of each agroindustrial by-product, it is important the study of these factors to develop and optimize the extraction method [44].

## 1.3.2   FAE

Polyphenols are secondary metabolites present in plants and are well-regarded as preventive agents of oxidative stress-related diseases, including cancer, cardiovascular diseases, diabetes, and neurodegenerative diseases if regularly consumed in the diet [45]. As mentioned above, conventional extraction methods of plant polyphenols (solvent extraction, Soxhlet extraction and hydrodistillation) are usually laborious and yield low extraction efficiencies. Therefore, solid-state fermentation (SSF) and submerged fermentation (SmF) are commonly used to liberate polyphenols from varied sources like creosote bush (*Larrea tridentata*) [46], tar bush [47], cashew husk [48], ferulic acid [49] and eugenol [50]. However, SmF is best used for enzyme production rather than extraction of polyphenols or other biomolecules.

### 1.3.2.1   SmF

SmF or liquid culture is defined as single cells or small cell aggregates in agitated liquid media [51].

### 1.3.2.1.1   Advantages and Disadvantages

Each extraction, biotransformation or production method of polyphenols may present advantages and disadvantages. For example, the culture medium composition is one of the most crucial factors to be considered if the product yield needs to be increased because the carbon and nitrogen sources and the carbon/nitrogen ratio are directly related to cell growth and metabolite biosynthesis rate [52]. Furthermore, the reproducibility is strongly affected by the oxygen mass transfer [53]. Table 1.7 shows a few of the most cited advantages and disadvantages of SmF.

Nevertheless, production of enzymes in SmF is still preferred by the industry, because all process parameters can be easily controlled. Several parameters, for instance, temperature, agitation, aeration, foam control and pH are dependent on the reactor type [53, 54]. Nevertheless, any process or method to liberate polyphenols presents advantages and disadvantages, and the choice will depend on the operational limitations, microbial performance and its enzyme production [55].

### 1.3.2.1.2   Process Parameters

The control of the process parameters is vital for the release of polyphenolic compounds. Some of the most important parameters are shown in Table 1.8.

**TABLE 1.7**  Advantages and Disadvantages of Submerged Fermentation

| Submerged Fermentation | |
|---|---|
| **Advantages** | **Disadvantages** |
| • Easy scale-up | • Is more used for enzyme production |
| • Easy purification of final products | • Works with low substrate ratio |
| • Easy biomass estimation | • The use of agricultural wastes streams and substrates may affect the reproducibility due to its heterogenic nature |
| • Control of aeration | |
| • Control of pH | |
| • Control of agitation | |
| • Improves the substrate accessibility | |
| • Easy control of temperature | |
| • The use of dissolvable C- and N-sources do not affect the reproducibility | |
| • Easy product recovery | |

Xu and Zhu [59] mentioned that the time of fermentation affects the rate of lignocellulose decomposition in corn stover, due to the exopolysaccharides synthesized by the fungi that accumulate around the mycelia, which restrict the interaction between the mycelia and corn stover. Also, the use of surfactants and their concentration level alter the microbial cell permeability and cell surface hydrophobicity, while too high a concentration and contact time, lead to microbial cell membrane disintegration, and these influence the mycelial biomass and polyphenol liberation [57, 63].

### 1.3.2.1.3  Use of Experimental Design Statistical Methodology

The fermentation conditions need to be optimized to maximize the liberation of the polyphenols. The conditions include the substrate pretreatment, the selection of microorganisms, pH, temperature, agitation, and mass/volume ratio, among others. Hence, experimental design tools have frequently been applied to reduce the number of experiments required and identify the best extraction conditions, considering all the factors studied. Some common experimental designs are Plackett-Burman [64, 65], central composite design [66, 67], Box-Behnken [68], principal component

**TABLE 1.8**  Process Parameters for Submerged-Fermentation Extraction of Polyphenols

| Parameters | Inoculum | Medium supplementation | Highlights | Ref. |
|---|---|---|---|---|
| 4 days; 28°C; pH 5.6; ratio 8% v/v; particle size of 60–100 mesh; 150 rpm. | Inonotus obliquus | Glucose, peptone, yeast extract, $KH_2PO_4$, $MgSO_4$, and $CaCl_2$ | Epigallocatechin–3-gallate, epicatechin–3-gallate, phelegridin G, davallialactone, and inoscavin B were liberated and phenolic acids were decreased by lignocellulose degradation | [56] |
| 9 days; 28°C; pH 5.6; ratio 9% v/v; 150 rpm | Inonotus obliquus | Glucose, peptone, yeast extract, $KH_2PO_4$, $MgSO_4$, and $CaCl_2$ | Ferulic acid and gallic acid, and flavonoids, i.e., epicatechin-3-gallate (ECG), epigallo-catechin-3-gallate (EGCG), and naringin were the main components. | [57] |
| 37°C; pH 4.5; ratio 10% v/v; particle size 0.1–1000 μm; 200–00 rpm; 20% $O_2$ saturation | Phanerochaete chrysosporium ATCC 24275 | Only apple pomace sludge | An increase in polyphenol content extracted by acetone (383–720 mg GAE/L) was observed during fermentation. | [58] |
| 28°C, pH 5.5; ratio 10% v/v; 150 rpm | Inonotus obliquus | Corn flour, peptone, $KH_2PO_4$, $ZnSO_4 \cdot 2H_2O$, $K_2HPO_4$, $FeSO_4 \cdot 7H_2O$, $MgSO_4 \cdot 7H_2O$, $CuSO_4 \cdot 5H_2O$, $CoCl_2$, and $MnSO_4 \cdot H_2O$ | Were liberated up to 135 mg of GAE of extracellular phenolic compounds from corn stover with higher antioxidant activity. | [59] |
| 28°C, 120 rpm | Aspergillus tamarii | Tannic acid | 0.36 g of Gallic acid per g of tannic acid was produced | [60] |
| 30°C, pH 6.0 | Pediococcus acidilactici | Rice bran | After hydrolysis of ferulic acid, 4-ethyl-phenol, vanillin, vanillic acid, and vanillyl alcohol were detected | [61] |
| 30°C; pH 6.5; liquid-solid ratio 12:1 (mL/g); 130 rpm | Polygonum cuspidatum | Roots flour | Polydatin conversion rate of 96.7% was reached | [62] |

analysis [69] and response surface methodology (RSM) [70, 71]. Several softwares are used for this task, such as Statistica, Minitab®, Statgraphics, JMP, and SPSS. In most instances, the aim is to obtain the best conditions for a statistical analysis that allows the reliability of the data collected, as well as the reproducibility and the scaling of the experiments.

### 1.3.2.2   SSF

SSF has been defined as any fermentation process, where the microorganisms grow on a solid or semisolid medium in the absence or near absence of free water in the substrate [72, 73]. However, the moisture content must be sufficient to allow microbial growth. These conditions are highly favorable for the growth of filamentous fungi, due to their capacity to grow at low free water levels and the ability for hyphal extension between the particle spaces of the solid matrices [74]. SSF has a high potential as a sustainable alternative for bio-based processes in which the substrates used are agroindustrial byproducts, such as bagasse, bran, straws, hulls, pulps, and peels, obtaining bio-products with high-added value, like enzymes [75, 76].

### 1.3.2.2.1   Advantages and Disadvantages

SSF has become a highly attractive alternative to traditional SmF, within the food and pharmaceutical industries [72]. Successful achievement of high titers of enzymes and organic acids by SSF has been documented [77, 78]. One of the primary advantages of SSF is the production of metabolites or enzymes by filamentous fungi, because the conventional substrates used for this purpose have similar composition and are exposed to comparable conditions as those found in the natural habitat of such fungi [76, 79]. Compared to liquid fermentation, SSF exhibits advantages and disadvantages (Table 1.9) [76, 78, 53, 82, 83]. The main disadvantages of SSF relative to SmF are associated with engineering aspects, and consequently, it is difficult to conduct a scale-up for industrial production [53]. Thus, the challenges that must be overcome are: (1) to ensure both heat and mass transfers within the packaging substrate, (2) to design tools for monitoring in-line parameters, such as pH and biomass, and (3) to provide the proper agitation system for the complete integration of substrates without damaging the microbial culture [75, 83].

**TABLE 1.9** Advantages and Disadvantages of SSF Compared with SmF

| Advantages | Disadvantages |
|---|---|
| • *Low energy requirements* | • *Minor homogeneity of culture medium* |
| Demand less energy for sterilization because of lower water activity. | SmF has higher homogeneity in the culture by the agitation process, which is easier than SmF. |
| • *Simplicity in substrates* | • *Biomass determination* |
| Solid support provides the nutrients, and commonly the substrates are agro-industrial wastes | Difficulties on separate biomass from solid substrates |
| • *Higher yield production* | • *Instrumentation and control* |
| The substrates are similar to natural habitat of microbial cultures; hence the microbial activity is increased. | Monitoring of parameters such as pH, temperature, dissolved O2 is easier in SmF as consequence of major homogeneity in the culture medium. |
| • *Friendly with the environment* | • *Pretreatments to the substrates* |
| Minimum water consumption and reduced number of effluents. | Sometimes the substrates must be ground. The substrates used in SmF are more simples |
| • *Minor risk of bacterial contamination* | • *Recuperation of products* |
| Bacterial cultures require high water activity than fungi cultures. Moreover, voluminous inoculum of fungi completes with bacterial inoculum | Recovery of extracellular enzymes by simple filtration or centrifugation. |
| • *Use of water-insoluble substrates* | • *Design of reactors* |
| The moist insoluble substrates provide carbon, nitrogen, minerals, etc. | Mixing, aeration, and scale-up is easier in SmF |

## 1.3.2.2.2 Process Parameters

Various factors affect the SSF performance. These factors can be varied according to the type of substrates, the microorganisms used and the scale of the process. Some of these factors are temperature, pH, package density, particle size, inoculum, aeration, agitation, medium supplementation, moisture, and water activity ($a_w$).

***Temperature:*** The optimal temperature of incubation during the fermentation process is determined by the microbial strain. Krishna [75] stated that the temperature is the most crucial parameter involved in SSF performance, but temperature control is difficult in SSF because of the low thermal conductivity. Moreover, the microbial growth promotes an increase in the temperature of the reactors, especially in the central zones of the reactor. The increment in temperature affects the microbial growth, spore germination and the yield of the product [73]. In general terms, the temperature modifies the growth rate, dissolved oxygen tension, rate of medium evaporation, pellet formation and product formation [53].

*pH:* The pH is a critical factor in both types of fermentation, SmF, and SSF. However, the monitoring and control of pH through SSF is not easy [79]. In SmF systems, all phases present are partially homogeneous, in contrast to the SSF process, where the three-phasic system (gas-liquid-solid) is highly heterogeneous, and the currently-available equipment and electrodes are not adequate [75]. For these reasons, the pH of the fermentation medium is adjusted at the beginning of the SSF. However, the pH value in the SSF is not constant, varying with the production of some metabolites produced by the microbial culture, such as organic acids [75, 79]. With the objective of controlling the variability of pH values during SSF, it is sometimes necessary to add buffers to the medium formulation.

*Package density:* Costa et al. [84] defined packing density as the ratio of the initial medium mass to the reactor volume occupied by the medium. In general, a high packing density results in limited heat and mass transfer [72]. Dilipkumar et al. [85] used copra waste for the SSF production of inulinase in batch and packed bed reactors. The authors observed a significant effect of packing density (30–50 g/L) on inulinase production and the optimum packaging value for maximum inulinase production was 38 g/L. The bed packing is related to the porosity of the material (that represents the intra- and inter-particle void spaces in the bioreactor bed) [75]. When packing density is too high, the air flow is limited and, therefore, the microbial growth is inhibited. Hence, it is worthwhile to find an optimum packing density that allows the metabolic activity of the microorganisms [85].

*Particle size:* The SSF process is constituted by three different physical phases (gas-liquid-solid) [86]. The aqueous phase is absorbed into the solid phase surface but is also in contact with the gaseous phase. At the liquid-solid interphase, the processes of mass and heat transfer are a result of substrate bioconservation [74, 87]. The liquid-solid interphase depends on the physical properties of the solid medium, including surface area, porosity, and mainly particle size [86]. Most substrates used as the carbon and energy source in SSF are agroindustrial wastes, and their particle size may affect the interspaces (porosity) and flow patterns in the solid substrate, affecting the rate of oxygen transfer, and, in turn, the microbial growth [88]. For these reasons, in some instances, the particle size of the substrate must be reduced by chopping or grinding, to render the nutrients and structure more accessible and susceptible to mycelial penetration [74, 87]. In general, substrates with particles that are too small may result in agglomeration, which causes poor microbial growth. Conversely, larger particles provide better airflow, but limited surface area available for microbial attack.

*Inoculum:* The type of inoculum (number of microorganisms initially present in the medium) depends on both the nature of the microorganism and the study objective [79]. For instance, Thanapimmetha et al. [89] used deoiled *Jatropha curcas* seed cake as a substrate for protease production by *Aspergillus oryzae* and observed an increment in protease production when the inoculums size was increased from 1% to 10%. Xiao et al. [90] optimized the tannase production from *Aspergillus tubingensis*, using tea stalks under SSF. The authors observed that inoculum size and incubation time had a significant effect on the tannase yield. Typically, an increase in the inoculum size increases the microbial growth. Nevertheless, if the inoculum level is too high, it could promote competition between micro-organisms and limit the substrate [89].

*Aeration and agitation:* Aeration and agitation are vital parameters in SSF due to affect the mass and heat transport phenomena to inter and intrapar-ticle level. Interparticle mass transfer refers to the movement of oxygen from the void fraction within the substrate to the growing microorganism, whereas, the passage of nutrients and enzymes within the substrate solid mass is known as intraparticle mass transfer [88].

The aeration rate during SSF depends on factors, such as the growth requirements of the microorganism, the production of gaseous and volatile metabolites, and heat evolution [75]. Sometimes, to counteract the incre-ments in temperature due to microbial metabolic activity, the aeration rate has to be increased. However, aeration could be difficult if the process is scale-up [76]. As previously mentioned [88], aeration promotes both oxygen flow and microbial growth. However, Prado et al. [91] observed that the citric acid production was favored by a low biomass development of *Aspergillus niger*, which occurred at low aeration rates during SSF.

*Medium supplementation:* In SSF, the solid material acts both, as a source of nutrients and physical support for microorganisms. In other cases, the solid material does not contain nutrients for microorganisms and only serves as a support for microbial growth. In the latter case, the nutrients must be added to the medium [74, 75]. Most of the agroindustrial byproducts used as substrates in SSF (e.g., sugarcane bagasse, wheat, rice, maize, and grain brans) supply the nutrients needed for the microbial growth. Nevertheless, sometimes supplementary nutrients are added to the substrates, with the objective of enhancing the product yield [92, 93]. Some nutrients, such as carbon, nitrogen, minerals, and vitamins, can regulate sporulation through metabolic effects [75]. Supplementary sources of carbon commonly used are sucrose, lactose, raffinose, maltose, cellobiose, malt extract, glycerol, and ethanol, among others [75, 92]. Nitrogen-containing sources include organic materials like yeast extract, peptone, soybean meal, soy peptone and urea. Inorganic compounds commonly used are ammonium sulfate, ammonium chloride, and potassium nitrate, for example [92]. Some minerals, such as $Na^+$, $Ca^{++}$, $Ni^+$, $Cu^+$, $Fe^{++}$, $Mn^{++}$, $K^+$, $Zn^{++}$, $Mg^{++}$ and $Mo^{++}$, improve fungal sporulation [75].

*Moisture and $a_w$:* High moisture levels promote a decrement in substrate porosity, which reduces the oxygen transfer and might enhance the risk of anaerobic bacteria contamination [73]. Conversely, when the moisture content is too low, it could affect the accessibility of nutrients, resulting in poor microbial growth. The substrate water content is related to the $a_w$ value. Poorna and Prema [94] verified that a reduced moisture content in the substrate decreases the swelling capacity of the substrate, which increases the water surface tension, reducing the $a_w$ for the microbial metabolic requirement. This parameter plays a key role in mass transfer (water and solutes) across the microbial cells [78]. Furthermore, the moisture level during SSF does not remain constant and must be monitored. Sometimes, it is necessary to add water directly to the system [79]. Typically, each type of microorganism that can grow in an SSF system requires a certain $a_w$ value. The $a_w$ requirements for different microorganisms in ascending order are filamentous fungi < yeasts < bacteria. The filamentous fungi are the ideal microorganisms to grow in an SSF system, although some bacterial species have been reported for the enzymes production by SSF [83]. The moisture levels described for SSF processes tend to vary between 30 and 85% [75].

### 1.3.2.2.3   *Use of Experimental Design Statistical Methodology*

The design of experiments (DOE) is a mathematical tool with well-established application in the biotechnology sector [80]. The optimization of parameters involved in SSF is a major priority that determines the economic feasibility of the process [95]. The classical one-factor-at-a-time statistical method to optimize medium fermentation and cultural conditions entails changing one variable (pH, temperature, agitation, etc.), while all other parameters are maintained constant [81, 93]. This approach is inconvenient because it requires many experimental data, deeming it time-consuming and expensive for a large number of variables [93, 94]. At present, there is an increasing interest in the use of statistical methods to facilitate identifying the optimal conditions for SSF process and, moreover, understand the interactions between the different physicochemical parameters (moisture, pH, agitation), using a minimum number of experiments [97]. The objective of the conventional experimental design is to find the best treatment among all the treatments evaluated, while RSM and Taguchi's design are more efficient and complete strategies determine the optimal process conditions. RSM DOE has been reported for the optimization of the SSF process [81]. The RSM DOE is a grouping of powerful statistical techniques useful for evaluating and modeling the effect of many variables on one variable of response, whose objective is to optimize this response [77]. In this regard, Berikten, and Kivanc [98] maximized the phytase production by *Thermomyces lanuginosus* in SSF by using RSM DOE. The authors registered an overall 10.83-fold enhancement in phytase activity due to the optimization. Conversely, the Taguchi DOE is a fractional factorial experiment design developed in 1950, for optimizing the products and processes in quality engineering. The objective of Taguchi's design model is to obtain the maximum information about the principal effect of the factors on a process with the minimal number of combination of treatments. In analysis of robust designs, Taguchi's method is based on the signal-to-noise ratio, which evaluates the quality as a function of variation, and orthogonal array, which accommodates many design factors simultaneously, to assess the effect of the maximal number of parameters at select levels with the minimal set of experiments [77, 80]. Compared to RSM DOE, Taguchi's method is more powerful for SSF optimization because it requires 50% less time than RSM [77, 96]. The Taguchi DOE was applied by Thanapimmetha et al. [89] to optimize the protease production by *A. oryzae* in SSF using *J. curcas* residue as substrate. The authors observed that the protease production increased up to 4.6 times compared to the non-optimized experiment. Another statistical tool used in SSF optimization is the Plackett–Burman DOE [81, 99, 100], which is useful for evaluating the interactions between parameters, as well as to decide the optimal interaction of variables for the response investigated [90]. Moreover, the Plackett-Burman DOE is a statistical tool that allows screening of key variables for further optimization in a rational way [101]. In this sense, Xiao et al. [90] used a sequential statistical strategy to optimize tannase production from *A. tubingensis* using tea stalks by SSF. First, by applying a Plackett-Burman DOE, the authors found that inoculum size and incubation time were the most significant factors for tannase yield. Next, a single steepest ascent experiment and central composite design with RSM DOE were applied. The optimization process increased the tannase yield up to two-fold.

## 1.4   APPLICATIONS OF EAE AND FAE

The ever-growing demand to extract bioactive compounds from byproducts has encouraged a continuous search for convenient extraction methods. Although some phytochemicals in the plant matrices are dispersed in the plant cell cytoplasm, others are bound within the polysaccharide-lignin network by hydrogen or hydrophobic bonds, which are not accessible with conventional solvent extraction [5, 17]. There are several alternatives for releasing these compounds of interest, such as EAE, and FAE.

EAE provides a safe, environmentally friendly and novel approach for the extraction of natural compounds known as bioactives. Several researchers have reported this method as the best choice for the extraction of bioactives from a variety of plant sources, like apple peel, citrus peel, grape pomace, berries, and oat bran [1, 102, 103]. For instance, the carotenoids from marigold flower or tomato peel, vanillin from vanilla green pods, and oil and polyphenols from grape seed have all been achieved using EAE [104–108].

The FAE is a biotechnological process that allows the re-use of agroindustrial byproducts, as a substrate support for biotechnological production into high value-added products, such as bioactive phenolic compounds. These fermentation techniques have been modified and refined to maximize productivity for economic and environmental advantages. Fermentation processes may be divided into SmF and SSF, which have led to the industrial-level production of bioactive compounds.

### 1.4.1 BIOACTIVE COMPOUNDS

The processing of fruit and vegetables produces enormous amounts of diverse byproducts, such as peels, seeds, stones, residual pulp and discarded whole pieces, which are rich in phenolic compounds, carotenoids, dietary fiber, vitamin C and minerals that might be a low-cost source to obtain functional ingredients [109].

Bioactive compounds attract considerable research interest due to their properties and benefits for human health. One of the major classes of bioactive compounds are phenolic compounds (also referred to as polyphenols) [110–112]. Phenolic compounds comprise flavonoids, phenolic acids and tannins, among others, and are present in all plant foods, but their type and levels vary enormously depending on the plant, genetic factors and environmental conditions [113]. Indeed, plant phenolics constitute the largest group of phenolic compounds, accounting for over half of the 8000 naturally occurring phenolic compounds [114]. Although many bioactive compounds are still produced by SmF, in the last decade, there has been an increasing trend towards the utilization of the SSF technique, as this process has been shown to be more efficient than SmF [115].

The polyphenols in pomegranate (*Punica granatum* L.) byproducts include anthocyanins (derived from delphinidin, cyanidin, pelargonidin), hydrolyzable tannins (catechin, epicatechin, punicalin, pedunculagin, punicalagin, gallic, and ellagic acid esters of glucose) [116, 117], as well as several lignans (isolariciresinol, medioresinol, matairesinol, pinoresinol, syringaresinol, secoisolariciresinol). Such constituents have been shown to reduce oxidative stress [118]. The byproducts of the olive industry are a major source of phenolic compounds. The phenolic compounds present in olives are distributed in olive oil and the wastewater in the aqueous phase or solid phase pomace, with only 1–2% partition in the former [119]. Winemaking generates a generous amount of byproducts rich in phenolic acids, flavonoids, including anthocyanins, as well as proanthocyanidins [120, 121]. Byproducts of the citrus industry possess a potential as being rich sources of bioactive compounds (ascorbic acid and flavonoids). It has been reported that the peel contains higher amounts of total phenolics compared to the edible portions [122, 123]. Cranberry pomace, the by-product of the cranberry juice processing industry, has also been highlighted as a good source of ellagic acid and other phenolic compounds [124]. The phenolic compounds in several other agroindustrial byproducts are illustrated in Table 1.10.

### 1.4.1.1 BIOLOGICAL PROPERTIES

Nowadays, phenolic compounds have attracted considerable interest due to their "protective" properties for human health, which are associated

**TABLE 1.10** Phenolic Compounds From Agricultural Byproducts

| By-product | Phenolic compounds | Ref. |
|---|---|---|
| Almond hulls | Chlorogenic acid, 3 and 4-O- Caffeoylquinic acid | [125] |
| Apple peels | Flavonoids, Anthocyanin | [126] |
| Artichoke blanching waters | Neochlorogenic acid, Cryptochlorogenic acid, Chlorogenic acid, Cynarin, Caffeic acid and derivates | [127] |
| Buckwheat hulls | Protocatechuic acid, 3,4-Dihydroxybenzaldehyde, Hyperin, Rutin, and Quercetin | [128] |
| Coffee silverskin | Caffeoylquinic, coumaroylquinic, and feruloylquinic acid | [129] |
| Dried apple pomace | Flavonols, Dihydrochalcones, and Hydroxycinnamates | [130] |
| Dried coconut husk | 4-Hydroxybenzoic acid and Ferulic acid | [131] |
| Grape pomace | Gallic acid, caffeic acid, syringic acid, catechin, and epicatechin | [132] |
| Onion solid waste | Quercetin 4-O-glucoside, Quercetin, Cyanidin 3-O-glucoside and Protocatechuic acid | [133] |
| Pomegranate peels | Punicalagin, Punicalin, Gallagic acid, Ellagic acid and derivates | [134] |

mainly with their antioxidant activity. Some of these properties are anti-atherogenic, anti-mutagenic, anti-thrombotic, anticancer, vasodilatory effects and diabetes mellitus [115]. Recent studies on the biological sources and properties of ellagic acid exemplified the antioxidant, anti-inflammatory, antiviral, antimicrobial, antimutagenic, antitumor, and anticancer activities among its bioactive properties [135, 136]. Ellagic acid is present in considerable amounts in cranberry, raspberry, and pomegranate fruits [124, 137–140]. Other compounds of interest are carotenoids, phenolic compounds and vitamins [141]. Stajčić et al. [110] reported that tomato byproducts should be considered as a promising source of valuable bioactive compounds for improving human nutrient supply and reducing the risks of diseases caused by oxidative damage, such as cancer.

### *1.4.1.1.1 Antioxidant*

It has been stated that the byproducts from several fruits are an inexpensive source to obtain antioxidants [142]. Studies on the free radical-scavenging properties of flavonoids allowed the characterization of the major phenolic components as antioxidants [143]. These compounds inhibit the initiation or propagation of oxidative chain reactions, donating hydrogen atoms and neutralizing free radicals through several mechanisms: reducing activity, free radical-scavenging, potential chelation of pro-oxidant metals and quenching of singlet oxygen [144].

Several methods are available that enable efficient extraction of antioxidants from plant byproducts for commercial applications [145]. The addition of specific enzymes, such as cellulases, α-amylases and pectinases, during extraction enhances recovery by degrading the plant cell wall and hydrolyzing the structural polysaccharides and lipid bodies [145, 146]. In a previous study, the use of commercial cellulases was observed to increase the release and recovery of antioxidant phenols, notably, anthocyanins, from black currant pomace [147].

The potential of SSF for the improvement of phenolic contents and antioxidant properties of various fruit pomaces has been evaluated by employing several microorganisms [148]. Soybean products fermented by SSF using *Trichoderma harzianum* showed a stronger antioxidant activity than the unfermented products, which was probably related to the higher contents of phenolic acids, flavonoids, and aglycone isoflavones, with more free hydroxyl groups, achieved during SSF [149]. SSF also improved the chemical composition and bioactivity of stale rice, using *Cordyceps sinensis* [150].

### 1.4.1.1.2 Antimicrobial

Several antimicrobial products have been developed over the years to control pathogens. However, the development of antimicrobial resistance and the relatively narrow spectrum of the current antimicrobials have had limited success and the microbial contamination of food still poses an important public health and economic challenge [151, 152]. Cranberry and its products have traditionally been used to treat urinary tract infections and disorders of the intestinal tract caused by *Escherichia coli* O157:H7 [153]. The ellagitannins and ellagic acid precursors are reported as antimicrobials due to their inhibitory effect on bacteria, fungi, and parasites [136].

Tannins and hydrolyzed tannins are known to exhibit inhibitory effects on microorganisms, either by sequestering metal ions critical for the microbial growth and metabolism or inhibiting essential functions of the bacterial membrane, such as ion channels and proteolytic activity [154]. Furthermore, ellagic acid might disrupt the salt tolerance and cellular homeostasis of *Vibrio parahaemolyticus*, by inhibiting some key ion channels in its membrane. This disruption may be fatal for the microorganism [155]. Conversely, it has been suggested that ellagic acid inhibits the growth of pathogens in humans, probably linking to proteins present in the bacteria wall, such as *Bacillus*, *Staphylococcus*, and *Salmonella* [156, 157].

The lipid-water interface and partially hydrophobic are involved in the mechanisms underlying the antimicrobial properties of phenolic compounds. Phenolic mobilizations occurring during the solid-state bioprocessing of pomace produce different hydrophobicities, depending on the duration of fungal bioprocessing [155]. The fruit extracts from pomegranate are rich in ellagitannins and ellagic acid. These bioactive compounds caused the inhibition of methicillin-resistant *Staphylococcus aureus* strains [158]. Cheng et al. [120] noted that wine residue extracts exhibited antibacterial and antifungal effects against Gram-negative (*E. coli*) and Gram-positive (*S. aureus*) bacteria, and fungi (*Candida albicans*).

### 1.4.1.1.3   Antimutagenic

Active oxygen and free radicals are associated with various physiological and pathological events, such as inflammation, immunization, aging, mutagenicity, and carcinogenicity [159]. Hochstein and Atallah [160] suggested that compounds possessing antioxidant activity can inhibit mutation and cancer because they can scavenge a free radical or induce antioxidant enzymes. Thus, the dietary habits are important for human health. The human diet contains a great variety of natural antimutagens and anticarcinogens, such as fibers, polyphenolic compounds, flavonoids, isoflavones, tocopherol, and ascorbic acid, among others [161–163].

Many plant polyphenols, such as ellagic acid, catechins, chlorogenic acid, and caffeic and ferulic acids, as well as their dietary sources, such as tea, have been shown to act as potent antimutagenic and anticarcinogenic agents [164, 165]. These compounds may function as blocking agents, and they can prevent the biotransformation of pre-mutagens into reactive metabolites by inhibiting metabolic activation or by scavenging reactive molecules [166]. As suppressing agents, they may modulate intracellular processes involved in DNA repair mechanisms [167]. Pomegranate peel contains substantial amounts of polyphenols, such as ellagic tannins, ellagic acid and gallic acid, which may be responsible for the antimutagenicity of the peel extract [163]. The antimutagenicity activity and antioxidant properties of phenolic compounds are likely to be associated with their capacity to inhibit the DNA damage caused by the presence of free radicals. However, the inhibition of mutagenesis, in particular, is generally not based on one specific mechanism [168]. Methanolic extracts from the byproducts of tomato (grape, cherry, bola, and saladette type) were evaluated for their phenolic content, and antioxidant and antimutagenic activities, with all samples showing positive antimutagenic activity by *Salmonella typhimurium* assay [169]. Furthermore, differences in the phenolic contents were detected, which depended on the tomato type and part of the fruit analyzed, revealing the best results for peel from the grape type [169]. In another study, Negi et al. [163] exhibited that pomegranate peel extracts decreased the sodium azide mutagenicity in two *S. typhimurium* strains (TA100 and TA1535) and the inhibition ranged from weak to potent, depending on the solvent extractant.

### 1.4.1.1.4    Anticancer

The anticarcinogens may inhibit one or more stages of the carcinogenic process and prevent or delay the formation of cancer [167]. Seeram et al. [170] identified important ellagitannin phytochemicals in various berry fruit extracts, which are characterized by their biological properties, such as anticancer (colon, prostate, and leukemia) activity. Vitamin C, flavonoids, and β-carotene are potential antioxidants that protect against oxidation of biomolecules, such as DNA, proteins, and lipid membranes, thereby reducing the risk of cancers, cataract, and cardiovascular diseases [171].

Stajčić et al. [110] proposed tomato byproducts should be considered as a promising source of valuable bioactive compounds for improving human nutrient supply and reducing the risks of diseases caused by oxidative damage, such as cancer. Another study indicated that structural factors would explain the antioxidant, antiproliferative, and anti-metastasis properties of some citrus flavonoids [115].

### 1.4.1.1.5    Antiviral

The antiviral activity of phenolic compounds against several viruses, such as human immunodeficiency virus (HIV), herpes simplex virus and influenza virus, is well documented in the literature [172–174]. Benavente-García et al. [175] reported that flavonoids contain health-related properties, including antiviral activities. Research shows that natural products possess a broad range of biological activities, such as antitumor and antiviral activities [176]. Flavonoid-rich extracts and fractions from the fruit have been found to inhibit acyclovir-resistant strains of herpes simplex virus type I [177, 178]. Several oligomeric hydrolyzable tannins inhibited the replication of HIV *in vitro* [179]. Extracts of *Mangifera indica* L. have been reported to possess antiviral activity against HIV [180]. Also, apple pomace extracts exhibited a reduction in the viral replication levels in human pathogens of considerable clinical interest, such as herpes simplex virus types 1 and 2.

### 1.5   CONCLUSION AND PERSPECTIVES

The valorization of agroindustrial byproducts has become an approach to harness material considered as waste and contribute to minimizing pollution. The development of biotechnological processes, such as EAE and FAE, has allowed the diversification of bioactive compounds with potential applications in the food, chemical, and pharmaceutical industries. These methods have demonstrated more advantages than disadvantages. However, the choice of extraction technique should consider the nature of the by-product and, particularly, infrastructure, and final objective. The agroindustry will continue to increase in the future in response to agricultural consumption demand and, consequently, the agroindustrial byproducts will also expand. This situation poses a challenge to develop or improve the biotechnological processes for efficient extraction of the bioactive compounds contained in these wastes.

## KEYWORDS

- agroindustrial byproducts
- extraction methods
- phenolic compounds
- valorization

## REFERENCES

1. Gómez-García, R., Martínez-Ávila, G. C. G., & Aguilar, C. N., (2012). Enzyme-assisted extraction of antioxidative phenolics from grape (*Vitis vinifera* L.) residues. *3 Biotech.*, *2*(4), 297–300.
2. Preston, T. R., (1986). Better utilization of crop residues and by-products in animal feeding: Research guidelines. 2. A practical manual for research workers. *FAO Animal Production and Health Paper*, *50*(2), 154–154.
3. Paredes, C., Bernal, M. P., Cegarra, J., & Roig, A., (2002). Bio-degradation of olive mill wastewater sludge by its co-composting with agricultural wastes. *Biores Technol.*, *85*(1), 1–8.
4. Wang, L., & Weller, C. L., (2006). Recent advances in extraction of nutraceuticals from plants. *Trends Food Sci. Technol.*, *17*(6), 300–312.
5. Azmir, J., Zaidul, I. S. M., Rahman, M. M., Sharif, K. M., Mohamed, A., Sahena, F., et al., (2013). Techniques for extraction of bioactive compounds from plant materials: A review. *J. Food Eng.*, *117*(4), 426–436.
6. Mussatto, S. I., Ballesteros, L. F., Martins, S., & Teixeira, J. A., (2012). Use of agroindustrial wastes in solid-state fermentation processes. In: Kuan-Yeow, S., & Xinxin, G., (eds.), *Industrial Waste* (pp. 121–140). InTech: Croatia.

7. El-Tayeb, T. S., Abdelhafez, A. A., Ali, S. H., & Ramadan, E. M., (2012). Effect of acid hydrolysis and fungal biotreatment on agro-industrial wastes for obtainment of free sugars for bioethanol production. *Braz. J. Microbiol.*, *43*(4), 1523–1535.

8. Romelle, F. D., Rani, A., & Manohar, R. S., (2016). Chemical composition of some selected fruit peels. *Eur. J. Food Sci. Technol.*, *4*(4), 12–21.

9. Rodríguez, C. S., (2008). Exploitation of biological wastes for the production of value-added products under solid-state fermentation conditions. *Biotechnol., J.*, *3*(7), 859–870.

10. Gaiker, C. T., (2004). *Handbook for the Prevention and Minimization of Waste and Valorization of By-Products in European Agro-Food Industries*. Awarenet. BI-223-04.

11. Bistanji, G., Hamadeh, S., Hassan, H., Tami, F., & Tannous, R., (2000). The potential of agro-industrial byproducts as feeds for livestock in Lebannon. *Livestock Research for Rural Development* [Online]. http://www.cipav.org.co/lrrd/lrrd12/3/bist123.

12. CEC *Burning Agricultural Waste: A Source of Dioxins* (2014), Commission of Environmental Cooperation: Montreal, Canada, p. 6.

13. Qdais, H. A., Abdulla, F., & Qrenawi, L., (2010). Solid waste landfills as a source of green energy: Case study of Al akeeder landfill. *Jordan J. Mech. Ind. Eng.*, *4*(1), 69–74.

14. Buekens, A., & Huang, H., (1998). Comparative evaluation of techniques for controlling the formation and emission of chlorinated dioxins/furans in municipal waste incineration. *J. Hazard. Mater.*, *62*(1), 1–33.

15. Deng, G. F., Shen, C., Xu, X. R., Kuang, R. D., Guo, Y. J., Zeng, L. S., et al., (2012). Potential of fruit wastes as natural resources of bioactive compounds. *Int. J. Mol. Sci.*, *13*(7), 8308.

16. Makris, D. P., Boskou, G., & Andrikopoulos, N. K., (2007). Recovery of antioxidant phenolics from white vinification solid by-products employing water/ethanol mixtures. *Biores Technol*, *98*(15), 2963–2967.

17. Das, A., Adsare, S., Das, M., & Kulthe, P. G., (2016). Chapter 2: Advanced techniques in extraction of phenolics from cereals, pulses, fruits, and vegetables. In: *Plant Secondary Metabolites* (Vol. 3, pp. 27–76). Canada: Apple Academic Press.

18. Ascacio-Valdés, J. A., Buenrostro-Figueroa, J. J., Aguilera-Carbó, A., Prado-Barragán, A., Rodríguez-Herrera, R., & Aguilar, C. N., (2011). Ellagitannins: Biosynthesis, biodegradation and biological properties *J. Med. Plants Res.*, *5*(19), 4696–4703.

19. Liu, J. J., Gasmalla, M. A. A., Li, P., & Yang, R., (2016). Enzyme-assisted extraction processing from oilseeds: Principle, processing and application. *Innov. Food Sci. Emerg. Technol.*, *35*, 184–193.

20. Chen, S., Xing, X. H., Huang, J. J., & Xu, M. S., (2011). Enzyme-assisted extraction of flavonoids from *Ginkgo biloba* leaves: Improvement effect of flavonol transglycosylation catalyzed by *Penicillium decumbens* cellulase. *Enzyme Microb. Technol.*, *48*(1), 100–105.

21. Sahne, F., Mohammadi, M., Najafpour, G. D., & Moghadamnia, A. A., (2017). Enzyme-assisted ionic liquid extraction of bioactive compound from turmeric (*Curcuma longa* L.): Isolation, purification and analysis of curcumin. *Ind. Crops Prod.*, *95*, 686–694.

22. Gai, Q. Y., Jiao, J., Wei, F. Y., Luo, M., Wang, W., Zu, Y. G., et al., (2013). Enzyme-assisted aqueous extraction of oil from *Forsythia suspense* seed and its physicochemical property and antioxidant activity. *Ind. Crops Prod., 51*, 274–278.

23. Hosni, K., Hassen, I., Chaâbane, H., Jemli, M., Dallali, S., Sebei, H., et al., (2013). Enzyme-assisted extraction of essential oils from thyme (*Thymus capitatus* L.) and rosemary (*Rosmarinus officinalis* L.): Impact on yield, chemical composition and antimicrobial activity. *Ind. Crops Prod., 47*, 291–299.

24. Kurmudle, N., Kagliwal, L. D., Bankar, S. B., & Singhal, R. S., (2013). Enzyme-assisted extraction for enhanced yields of turmeric oleoresin and its constituents. *Food Biosci., 3*, 36–41.

25. Lin, J. A., Kuo, C. H., Chen, B. Y., Li, Y., Liu, Y. C., Chen, J. H., et al., (2016). A novel enzyme-assisted ultrasonic approach for highly efficient extraction of resveratrol from *Polygonum cuspidatum. Ultrasons Sonochem., 32*, 258–264.

26. Liu, T., Sui, X., Li, L., Zhang, J., Liang, X., Li, W., et al., (2016). Application of ionic liquids based enzyme-assisted extraction of chlorogenic acid from *Eucommia ulmoides* leaves. *Anal. Chim. Acta., 903*, 91–99.

27. Basegmez, H. I. O., Povilaitis, D., Kitrytė, V., Kraujalienė, V., Šulniūtė, V., Alasalvar, C., et al., (2017). Biorefining of blackcurrant pomace into high value functional ingredients using supercritical $CO_2$, pressurized liquid and enzyme assisted extractions. *J. Supercrit Fluids, 124*, 10–19.

28. Mackėla, I., Andriekus, T., & Venskutonis, P. R., (2017). Biorefining of buckwheat (*Fagopyrum esculentum*) hulls by using supercritical fluid, soxhlet, pressurized liquid and enzyme-assisted extraction methods. *J. Food Eng, 213*, 38–46.

29. Boulila, A., Hassen, I., Haouari, L., Mejri, F., Amor, I. B., Casabianca, H., et al., (2015). Enzyme-assisted extraction of bioactive compounds from bay leaves (*Laurus nobilis* L.). *Ind. Crops Prod., 74*, 485–493.

30. De Camargo, A. C., Regitano-d'Arce, M. A. B., Biasoto, A. C. T., & Shahidi, F., (2016). Enzyme-assisted extraction of phenolics from winemaking by-products: Antioxidant potential and inhibition of alpha-glucosidase and lipase activities. *Food Chem., 212*, 395–402.

31. Nagendra, C. K. L., Manasa, D., Srinivas, P., & Sowbhagya, H. B., (2013). Enzyme-assisted extraction of bioactive compounds from ginger (*Zingiber officinale* Roscoe). *Food Chem., 139*(1–4), 509–514.

32. Kammerer, D., Claus, A., Schieber, A., & Carle, R., (2005). A novel process for the recovery of polyphenols from grape (*Vitis vinifera* L.) Pomace. *J. Food Sci., 70*(2), 157–163.

33. Li, B. B., Smith, B., & Hossain, M. M., (2006). Extraction of phenolics from citrus peels: II. Enzyme-assisted extraction method. *Sep. Purif. Technol., 48*(2), 189–196.

34. Kim, Y. J., Kim, D. O., Chun, O. K., Shin, D. H., Jung, H., Lee, C. Y., et al., (2005). Phenolic extraction from apple peel by cellulases from *Thermobifida fusca. J. Agric. Food Chem., 53*(24), 9560–9565.

35. Puri, M., Sharma, D., & Barrow, C. J., (2012). Enzyme-assisted extraction of bioactives from plants. *Trends in Biotechnol., 30*(1), 37–44.

36. Maier, T., Göppert, A., Kammerer, D. R., Schieber, A., & Carle, R., (2008). Optimization of a process for enzyme-assisted pigment extraction from grape (*Vitis vinifera* L.) pomace. *Eur. Food Res. Technol.*, *227*(1), 267–275.

37. Dinkova, R., Heffels, P., Shikov, V., Weber, F., Schieber, A., & Mihalev, K., (2014). Effect of enzyme-assisted extraction on the chilled storage stability of bilberry (*Vaccinium myrtillus* L.) anthocyanins in skin extracts and freshly pressed juices. *Food Res. Int.*, *65, Part A*, 35–41.

38. Dutta, S., & Bhattacharjee, P., (2015). Enzyme-assisted supercritical carbon dioxide extraction of black pepper oleoresin for enhanced yield of piperine-rich extract. *J. Biosci. Bioeng.*, *120*(1), 17–23.

39. Strati, I. F., Gogou, E., & Oreopoulou, V., (2015). Enzyme and high pressure assisted extraction of carotenoids from tomato waste. *Food Bioprod. Proc.*, *94*, 668–674.

40. Mushtaq, M., Sultana, B., Anwar, F., Adnan, A., & Rizvi, S. S. H., (2015). Enzyme-assisted supercritical fluid extraction of phenolic antioxidants from pomegranate peel. *J. Supercrit. Fluids.*, *104*, 122–131.

41. Zhou, Z., Shao, H., Han, X., Wang, K., Gong, C., & Yang, X., (2017). The extraction efficiency enhancement of polyphenols from *Ulmus pumila* L. barks by trienzyme-assisted extraction. *Ind. Crops Prod.*, *97*, 401–408.

42. Zuorro, A., Fidaleo, M., & Lavecchia, R., (2011). Enzyme-assisted extraction of lycopene from tomato processing waste. *Enzyme Microb. Technol.*, *49*(6–7), 567–573.

43. Bhanja, D. T., Chakraborty, S., Jain, K. K., Sharma, A., & Kuhad, R. C., (2016). Antioxidant phenolics and their microbial production by submerged and solid-state fermentation process: A review. *Trends Food Sci. Technol.*, *53*, 60–74.

44. Zhang, L., Tu, Z. C., Wang, H., Fu, Z. F., Wen, Q. H., Chang, H. X., et al., (2015). Comparison of different methods for extracting polyphenols from *Ipomoea batatas* leaves, and identification of antioxidant constituents by HPLC-QTOF-MS2. *Food Res. Int.*, *70*, 101–109.

45. Ng, H. S., Teoh, A. N., Lim, J. C. W., Tan, J. S., Wan, P. K., Yim, H. S., Show, P. L., & Lan, J. C. W. Thermo-sensitive aqueous biphasic extraction of polyphenols from *Camellia sinensis* var. assamica leaves. *J. Taiwan Inst. Chem. Eng, 79*, 151–157.

46. Treviño-Cueto, B., Luis, M., Contreras-Esquivel, J. C., Rodríguez, R., Aguilera, A., & Aguilar, C. N., (2007). Gallic acid and tannase accumulation during fungal solid-state culture of a tannin-rich desert plant (*Larrea tridentata* Cov.). *Biores. Technol.*, *98*(3), 721–724.

47. Ventura, J., Gutiérrez-Sanchez, G., Rodríguez-Herrera, R., & Aguilar, C. N., (2009). Fungal cultures of tar bush and creosote bush for production of two phenolic antioxidants (Pyrocatechol and Gallic acid). *Folia Microbiol.*, *54*(3), 199–203.

48. Lokeswari, N., Sriramireddy, D., Sudhakararao, P., & Varaprasad, B., (2010). Production of gallic acid using mutant strain of *Aspergillus oryzae*. *J. Pharm. Res.*, *3*(6), 1402–1406.

49. Huang, Y. C., Chen, Y. F., Chen, C. Y., Chen, W. L., Ciou, Y. P., Liu, W. H., et al., (2011). Production of ferulic acid from lignocellulolytic agricultural biomass by *Thermobifida fusca* thermostable esterase produced in *Yarrowia lipolytica* transformant. *Biores. Technol.*, *102*(17), 8117–8122.

50. Lambert, F., Zucca, J., Ness, F., & Aigle, M., (2014). Production of ferulic acid and coniferyl alcohol by conversion of eugenol using a recombinant strain of *Saccharomyces cerevisiae*. *Flavor Frag. J.*, *29*(1), 14–21.

51. Preil, W., (2005). General introduction: A personal reflection on the use of liquid media for in vitro culture. In: Hvoslef-Eide, A. K., & Preil, W., (eds.), *Liquid Culture Systems for In Vitro Plant Propagation* (pp. 1–18). Springer Netherlands: Dordrecht.

52. Fu, L., Wang, J., Xu, S., Hao, L., & Wang, Y., (2014). Optimization of submerged culture for exopolysaccharides production by *Morcella esculenta* and its antioxidant activities *in vitro*. In: Zhang, T. C., Ouyang, P., Kaplan, S., & Skarnes, B., (eds.), *Proceedings of the 2012 International Conference on Applied Biotechnology (ICAB 2012)* (Vol. 1, pp. 535–545). Springer Berlin Heidelberg: Berlin, Heidelberg.

53. Hansen, G. H., Lübeck, M., Frisvad, J. C., Lübeck, P. S., & Andersen, B., (2015). Production of cellulolytic enzymes from ascomycetes: Comparison of solid-state and submerged fermentation. *Proc. Biochem.*, *50*(9), 1327–1341.

54. Vaidyanathan, S., Macaloney, G., Vaughan, J., McNeil, B., & Harvey, L. M., (1999). Monitoring of submerged bioprocesses. *Crit. Rev. Biotechnol.*, *19*(4), 277–316.

55. Alves-Prado, H. F., Gomes, E., & Da Silva, R., (2006). Evaluation of solid and submerged fermentations for the production of cyclodextrin glycosyltransferase by *Paenibacillus campinasensis* H69-3 and characterization of crude enzyme. In: McMillan, J. D., Adney, W. S., Mielenz, J. R., & Klasson, K. T., (eds.), *Twenty-Seventh Symposium on Biotechnology for Fuels and Chemicals* (pp. 234–246). Humana Press: Totowa, NJ.

56. Xu, X. Q., Hu, Y., & Zhu, L. H., (2014). The capability of *Inonotus obliquus* for lignocellulosic biomass degradation in peanut shell and for simultaneous production of bioactive polysaccharides and polyphenols in submerged fermentation. *J. Taiwan Inst. Chem. Eng.*, *45*(6), 2851–2858.

57. Xu, X., Zhao, W., & Shen, M., (2016). Antioxidant activity of liquid cultured *Inonotus obliquus* polyphenols using tween-20 as a stimulatory agent: Correlation of the activity and the phenolic profiles. *J. Taiwan Inst. Chem. Eng.*, *69*, 41–47.

58. Gassara, F., Ajila, C. M., Brar, S. K., Verma, M., Tyagi, R. D., & Valero, J. R., (2012). Liquid state fermentation of apple pomace sludge for the production of ligninolytic enzymes and liberation of polyphenolic compounds. *Proc. Biochem.*, *47*(6), 999–1004.

59. Xu, X., & Zhu, J., (2011). Enhanced phenolic antioxidants production in submerged cultures of *Inonotus obliquus* in a ground corn stover medium. *Biochem. Eng. J.*, *58*, 103–109.

60. Costa, A. M. D., Souza, C. G. M. D., Bracht, A., Kimiko, M., Kadowaki, S. A. C. D. S. D., Oliveira, R. F., et al., (2013). Production of tannase and gallic acid by *Aspergillus tamarii* in submerged and solid-state cultures. *Afr. J. Biochem. Res.*, *7*(10), 197–202.

61. Kaur, B., Chakraborty, D., Kaur, G., & Kaur, G., (2013). Biotransformation of rice bran to ferulic acid by pediococcal isolates. *Appl. Biochem. Biotechnol.*, *170*(4), 854–867.

62. Jin, S., Luo, M., Wang, W., Zhao, C. J., Gu, C. B., Li, C. Y., et al., (2013). Biotransformation of polydatin to resveratrol in *Polygonum cuspidatum* roots by highly immobilized edible *Aspergillus niger* and Yeast. *Biores. Technol.*, *136*, 766–770.

63. Zhang, D., Zhu, L., & Li, F., (2013). Influences and mechanisms of surfactants on pyrene biodegradation based on interactions of surfactant with a *Klebsiella oxytoca* strain. *Biores. Technol., 142*, 454–461.

64. Jovanović, A. A., Đorđević, V. B., Zdunić, G. M., Pljevljakušić, D. S., Šavikin, K. P., Gođevac, D. M., et al., (2017). Optimization of the extraction process of polyphenols from *Thymus serpyllum* L. herb using maceration, heat- and ultrasound-assisted techniques. *Sep. Purif. Technol., 179*, 369–380.

65. Ćujić, N., Šavikin, K., Janković, T., Pljevljakušić, D., Zdunić, G., & Ibrić, S., (2016). Optimization of polyphenols extraction from dried chokeberry using maceration as traditional technique. *Food Chem., 194*, 135–142.

66. Zuorro, A., (2015). Optimization of polyphenol recovery from espresso coffee residues using factorial design and response surface methodology. *Sep. Purif. Technol., 152*, 64–69.

67. Karakashov, B., Grigorakis, S., Loupassaki, S., & Makris, D. P., (2015). Optimisation of polyphenol extraction from *Hypericum perforatum* (St. John's Wort) using aqueous glycerol and response surface methodology. *J. Appl. Res. Med. Aromat. Plants, 2*(1), 1–8.

68. Monsanto, M., Trifunovic, O., Bongers, P., Meuldijk, J., & Zondervan, E., (2014). Black tea cream effect on polyphenols optimization using statistical analysis. *Comput. Chem. Eng., 66*, 12–21.

69. Aires, A., Carvalho, R., & Saavedra, M. J., (2016). Valorization of solid wastes from chestnut industry processing: Extraction and optimization of polyphenols, tannins and ellagitannins and its potential for adhesives, cosmetic and pharmaceutical industry. *Waste Manag., 48*, 457–464.

70. Bouras, M., Chadni, M., Barba, F. J., Grimi, N., Bals, O., & Vorobiev, E., (2015). Optimization of microwave-assisted extraction of polyphenols from *Quercus* bark. *Ind. Crops Prod., 77*, 590–601.

71. Brahim, M., Gambier, F., & Brosse, N., (2014). Optimization of polyphenols extraction from grape residues in water medium. *Ind. Crops Prod., 52*, 18–22.

72. Chen, H., (2013). *Modern Solid-state Fermentation: Theory and Practice*. Springer: New York, 324 p.

73. Pandey, A., (1992). Recent process developments in solid-state fermentation. *Process Biochem., 27*(2), 109–117.

74. Ruiz, H. A., Rodríguez-Jasso, R. M., Rodríguez, R., Contreras-Esquivel, J. C., & Aguilar, C. N., (2012). Pectinase production from lemon peel pomace as support and carbon source in solid-state fermentation column-tray bioreactor. *Biochem. Eng. J., 65*, 90–95.

75. Krishna, C., (2005). Solid-state fermentation systems-an overview. *Crit. Rev. Biotechnol., 25*(1–2), 1–30.

76. Hölker, U., & Lenz, J., (2005). Solid-state fermentation - are there any biotechnological advantages? *Curr. Opin. Microbiol., 8*(3), 301–306.

77. Aggarwal, A., Singh, H., Kumar, P., & Singh, M., (2008). Optimizing power consumption for CNC turned parts using response surface methodology and Taguchi's technique-A comparative analysis. *J. Mater Process Technol., 200*(1), 373–384.

78. Pandey, A., (2003). Solid-state fermentation. *Biochem. Eng. J., 13*(2), 81–84.

79. Yoon, L. W., Ang, T. N., Ngoh, G. C., & Chua, A. S. M., (2014). Fungal solid-state fermentation and various methods of enhancement in cellulase production. *Biomass Bioenergy.*, *67*, 319–338.

80. Shehata, A. N., & Abd El Aty, A. A., (2014). Optimization of process parameters by statistical experimental designs for the production of naringinase enzyme by marine fungi. *Int. J. Chem. Eng.*, *2014*, 10.

81. Colla, L. M., Primaz, A. L., Benedetti, S., Loss, R. A., De Lima, M., Reinehr, C. O., et al., (2016). Surface response methodology for the optimization of lipase production under submerged fermentation by filamentous fungi. *Braz. J. Microbiol.*, *47*(2), 461–467.

82. Prado, B. L. A., Figueroa, J. J. B., Rodríguez, D. L. V., Aguilar, G. C. N., & Hennigs, C., (2016). Fermentative production methods. In: Poltronieri, P., & D'Urso, O. F., (eds.), *Biotransformation of Agricultural Waste and By-Products* (pp. 189–217). Elsevier: Amsterdam, Netherlands.

83. Singhania, R. R., Patel, A. K., Soccol, C. R., & Pandey, A., (2009). Recent advances in solid-state fermentation. *Biochem. Eng. J.*, *44*(1), 13–18.

84. Costa, J. A. V., Alegre, R. M., & Hasan, S. D. M., (1998). Packing density and thermal conductivity determination for rice bran solid-state fermentation. *Biotechnol. Tech.*, *12*(10), 747–750.

85. Dilipkumar, M., Rajasimman, M., & Rajamohan, N., (2014). Utilization of copra waste for the solid-state fermentative production of inulinase in batch and packed bed reactors. *Carbohydr. Polym.*, *102*, 662–668.

86. Membrillo, I., Sánchez, C., Meneses, M., Favela, E., & Loera, O., (2011). Particle geometry affects differentially substrate composition and enzyme profiles by *Pleurotus ostreatus* growing on sugar cane bagasse. *Bioresour. Technol.*, *102*(2), 1581–1586.

87. Bari, M., Alam, M., Muyibi, S. A., Jamal, P., & Mamun, A., (2010). Effect of particle size on production of citric acid from oil palm empty fruit bunches as new substrate by wild *Aspergillus niger*. *J. Appl. Polym. Sci.*, *10*(21), 2648–2652.

88. Raghavarao, K. S. M. S., Ranganathan, T. V., & Karanth, N. G., (2003). Some engineering aspects of solid-state fermentation. *Biochem. Eng. J.*, *13*(2), 127–135.

89. Thanapimmetha, A., Luadsongkram, A., Titapiwatanakun, B., & Srinophakun, P., (2012). Value added waste of *Jatropha curcas* residue: Optimization of protease production in solid-state fermentation by Taguchi DOE methodology. *Ind. Crops Prod.*, *37*(1), 1–5.

90. Xiao, A., Huang, Y., Ni, H., Cai, H., & Yang, Q., (2015). Statistical optimization for tannase production by *Aspergillus tubingensis* in solid-state fermentation using tea stalks. *Electron J. Biotechnol.*, *18*(3), 143–147.

91. Prado, F. C., Vandenberghe, L. P. S., Lisboa, C., Paca, J., Pandey, A., & Soccol, C. R., (2004). Relation between citric acid production and respiration rate of *Aspergillus niger* in solid-state fermentation. *Eng. Life Sci.*, *4*(2), 179–186.

92. Gajdhane, S. B., Bhagwat, P. K., & Dandge, P. B., (2016). Response surface methodology-based optimization of production media and purification of α-galactosidase in solid-state fermentation by *Fusarium moniliforme* NCIM 1099. *3 Biotech.*, *6*(2), 260.

93. Adinarayana, K., Ellaiah, P., Srinivasulu, B., Bhavani Devi, R., & Adinarayana, G., (2003). Response surface methodological approach to optimize the nutritional param-

eters for neomycin production by *Streptomyces marinensis* under solid-state fermentation. *Process Biochem.*, *38*(11), 1565–1572.

94. Asha, P. C., & Prema, P., (2007). Production of cellulase-free endoxylanase from novel alkalophilic thermotolerent *Bacillus pumilus* by solid-state fermentation and its application in wastepaper recycling. *Bioresour. Technol.*, *98*(3), 485–490.

95. Dogan, N. M., Sensoy, T., Doganli, G. A., Bozbeyoglu, N. N., Arar, D., Akdogan, H. A., & Canpolat, M., (2016). Immobilization of *Lycinibacillus fusiformis* B26 cells in different matrices for use in turquoise blue HFG decolourization. *Arch. Environ. Prot.*, *42*(2), 92.

96. Nandal, P., Ravella, S. R., & Kuhad, R. C., (2013). Laccase production by *Coriolopsis caperata* RCK2011: Optimization under solid-state fermentation by Taguchi DOE methodology. *Sci. Rep.*, *3*, 1386.

97. Hajji, M., Rebai, A., Gharsallah, N., & Nasri, M., (2008). Optimization of alkaline protease production by *Aspergillus clavatus* ES1 in *Mirabilis jalapa* tuber powder using statistical experimental design. *Appl. Microbiol. Biotechnol.*, *79*(6), 915.

98. Berikten, D., & Kivanc, M., (2014). Optimization of solid-state fermentation for phytase production by *Thermomyces lanuginosus* using response surface methodology. *Prep. Biochem. Biotechnol.*, *44*(8), 834–848.

99. Mazzucotelli, C. A., Moreira, M. D. R., & Ansorena, M. R., (2015). Statistical optimization of medium components and physicochemical parameters to simultaneously enhance bacterial growth and esterase production by *Bacillus thuringiensis*. *Can. J. Microbiol.*, *62*(1), 24–34.

100. Ghanem, N. B., Yusef, H. H., & Mahrouse, H. K., (2000). Production of *Aspergillus terreus* xylanase in solid-state cultures: application of the Plackett–Burman experimental design to evaluate nutritional requirements. *Bioresour. Technol.*, *73*(2), 113–121.

101. Rajendran, A., Palanisamy, A., & Thangavelu, V., (2008). Evaluation of medium components by Plackett-Burman statistical design for lipase production by *Candida rugosa* and kinetic modeling. *Chin. J. Biotech.*, *24*(3), 436–444.

102. Cerda, A., Martínez, M. E., Soto, C., Poirrier, P., Perez-Correa, J. R., Vergara-Salinas, J. R., et al., (2013). The enhancement of antioxidant compounds extracted from *Thymus vulgaris* using enzymes and the effect of extracting solvent. *Food Chem.*, *139*(1), 138–143.

103. Alrahmany, R., & Tsopmo, A., (2012). Role of carbohydrases on the release of reducing sugar, total phenolics and on antioxidant properties of oat bran. *Food Chem.*, *132*(1), 413–418.

104. Dehghan-Shoar, Z., Hardacre, A. K., Meerdink, G., & Brennan, C. S., (2011). Lycopene extraction from extruded products containing tomato skin. *Int. J. Food Sci. Technol.*, *46*(2), 365–371.

105. Yang, Y. C., Li, J., Zu, Y. G., Fu, Y. J., Luo, M., Wu, N., et al., (2010). Optimisation of microwave-assisted enzymatic extraction of corilagin and geraniin from *Geranium sibiricum* Linne and evaluation of antioxidant activity. *Food Chem.*, *122*(1), 373–380.

106. Passos, C. P., Yilmaz, S., Silva, C. M., & Coimbra, M. A., (2009). Enhancement of grape seed oil extraction using a cell wall degrading enzyme cocktail. *Food Chem.*, *115*(1), 48–53.

107. Barzana, E., Rubio, D., Santamaria, R. I., Garcia-Correa, O., Garcia, F., Ridaura Sanz, V. E., et al., (2002). Enzyme-mediated solvent extraction of carotenoids from marigold flower (*Tagetes erecta*). *J. Agric. Food Chem.*, *50*(16), 4491–4496.

108. Ruiz-Terán, F., Perez-Amador, I., & López-Munguia, A., (2001). Enzymatic extraction and transformation of glucovanillin to vanillin from vanilla green pods. *J. Agric. Food Chem.*, *49*(11), 5207–5209.

109. De Ancos, B., Colina-Coca, C., González-Peña, D., & Sánchez-Moreno, C., (2015). Bioactive compounds from vegetable and fruit by-products. In: Gupta, V. K., Tuohy, M. G., O'Donovan, A., & Lohani, M., (eds.), *Biotechnology of Bioactive Compounds, Sources and Applications, Section I* (pp. 3–36). John Wiley & Sons, Ltd, Chichester, UK.

110. Stajčić, S., Ćetković, G., Čanadanović-Brunet, J., Djilas, S., Mandić, A., & Četojević-Simin, D., (2015). Tomato waste: Carotenoids content, antioxidant and cell growth activities. *Food Chem.*, *172*, 225–232.

111. Correia, R. T., Borges, K. C., Medeiros, M. F., & Genovese, M. I., (2012). Bioactive compounds and phenolic-linked functionality of powdered tropical fruit residues. *Revista de Agroquímica y Tecnología de Alimentos*, *18*(6), 539–547.

112. Yu, J., & Ahmedna, M., (2013). Functional components of grape pomace: Their composition, biological properties and potential applications. *Int. J. Food Sci. Technol.*, *48*(2), 221–237.

113. Kris-Etherton, P. M., Hecker, K. D., Bonanome, A., Coval, S. M., Binkoski, A. E., Hilpert, K. F., et al., (2002). Bioactive compounds in foods: Their role in the prevention of cardiovascular disease and cancer. *Am. J. Med.*, *113*(9), 71–88.

114. Baxter, H., Harborne, J. B., & Moss, G. P., (1998). *Phytochemical Dictionary: A Handbook of Bioactive Compounds From Plants* (2nd edn.). CRC Press, Taylor & Francis, London, 976 p.

115. Martins, S., Mussatto, S. I., Martínez-Ávila, G., Montañez-Saenz, J., Aguilar, C. N., & Teixeira, J. A., (2011). Bioactive phenolic compounds: Production and extraction by solid-state fermentation. A review. *Biotechnol. Adv.*, *29*(3), 365–373.

116. Cuccioloni, M., Mozzicafreddo, M., Sparapani, L., Spina, M., Eleuteri, A. M., Fioretti, E., et al., (2009). Pomegranate fruit components modulate human thrombin. *Fitoterapia*, *80*(5), 301–305.

117. Gil, M. I., Tomás-Barberán, F. A., Hess-Pierce, B., Holcroft, D. M., & Kader, A. A., (2000). Antioxidant activity of pomegranate juice and its relationship with phenolic composition and processing. *J. Agric. Food Chem.*, *48*(10), 4581–4589.

118. Bonzanini, F., Bruni, R., Palla, G., Serlataite, N., & Caligiani, A., (2009). Identification and distribution of lignans in *Punica granatum* L. fruit endocarp, pulp, seeds, wood knots and commercial juices by GC–MS. *Food Chem.*, *117*(4), 745–749.

119. Rodis, P. S., Karathanos, V. T., & Mantzavinou, A., (2002). Partitioning of olive oil antioxidants between oil and water phases. *J. Agric. Food Chem.*, *50*(3), 596–601.

120. Cheng, V. J., Bekhit, A. E. D. A., McConnell, M., Mros, S., & Zhao, J., (2012). Effect of extraction solvent, waste fraction and grape variety on the antimicrobial and antioxidant activities of extracts from wine residue from cool climate. *Food Chem.*, *134*(1), 474–482.

121. De Camargo, A. C., Regitano-d'Arce, M. A. B., Biasoto, A. C. T., & Shahidi, F., (2014). Low molecular weight phenolics of grape juice and winemaking byproducts: Antioxidant activities and inhibition of oxidation of human low-density lipoprotein cholesterol and DNA strand breakage. *J. Agric. Food Chem.*, *62*(50), 12159–12171.

122. Lario, Y., Sendra, E., Garcı, X., A-Pérez, J., Fuentes, C., Sayas-Barberá, E., et al., (2004). Preparation of high dietary fiber powder from lemon juice by-products. *Innov. Food Sci. Emerg. Tech.*, *5*(1), 113–117.

123. Balasundram, N., Sundram, K., & Samman, S., (2006). Phenolic compounds in plants and agri-industrial by-products: Antioxidant activity, occurrence, and potential uses. *Food Chem.*, *99*(1), 191–203.

124. Vattem, D. A., & Shetty, K., (2003). Ellagic acid production and phenolic antioxidant activity in cranberry pomace *(Vaccinium macrocarpon)* mediated by *Lentinus edodes* using a solid-state system. *Process Biochem.*, *39*(3), 367–379.

125. Takeoka, G. R., & Dao, L. T., (2003). Antioxidant constituents of almond [*Prunus dulcis* (Mill.) D.A. Webb] hulls. *J. Agric. Food Chem.*, *51*(2), 496–501.

126. Wolfe, K. L., & Liu, R. H., (2003). Apple peels as a value-added food ingredient. *J. Agric. Food Chem.*, *51*(6), 1676–1683.

127. Llorach, R., Espín, J. C., Tomás-Barberán, F. A., & Ferreres, F., (2002). Artichoke *(Cynara scolymus* L.) Byproducts as a potential source of health-promoting antioxidant phenolics. *J. Agric. Food Chem.*, *50*(12), 3458–3464.

128. Watanabe, M., Ohshita, Y., & Tsushida, T., (1997). Antioxidant compounds from buckwheat (*Fagopyrum esculentum* Möench) hulls. *J. Agric. Food Chem.*, *45*(4), 1039–1044.

129. Bresciani, L., Calani, L., Bruni, R., Brighenti, F., Del Rio, D., (2014). Phenolic composition, caffeine content and antioxidant capacity of coffee silverskin. *Food Res. Int.*, *61*, 196–201.

130. Schieber, A., Hilt, P., Streker, P., Endreß, H. U., Rentschler, C., & Carle, R., (2003). A new process for the combined recovery of pectin and phenolic compounds from apple pomace. *Innov. Food Sci. Emerg. Tech.*, *4*(1), 99–107.

131. Dey, G., Sachan, A., Ghosh, S., & Mitra, A., (2003). Detection of major phenolic acids from dried mesocarpic husk of mature coconut by thin layer chromatography. *Ind. Crops Prod.*, *18*(2), 171–176.

132. Tournour, H. H., Segundo, M. A., Magalhães, L. M., Barreiros, L., Queiroz, J., & Cunha, L. M., (2015). Valorization of grape pomace: Extraction of bioactive phenolics with antioxidant properties. *Ind. Crops Prod.*, *74*, 397–406.

133. Katsampa, P., Valsamedou, E., Grigorakis, S., & Makris, D. P., (2015). A green ultrasound-assisted extraction process for the recovery of antioxidant polyphenols and pigments from onion solid wastes using Box–Behnken experimental design and kinetics. *Ind. Crops Prod.*, *77*, 535–543.

134. Ascacio-Valdés, J. A., Aguilera-Carbó, A. F., Buenrostro, J. J., Prado-Barragán, A., Rodríguez-Herrera, R., & Aguilar, C. N., (2016). The complete biodegradation pathway of ellagitannins by *Aspergillus niger* in solid-state fermentation. *J. Basic Microbiol.*, *56*(4), 329–336.

135. Buenrostro-Figueroa, J., Ascacio-Valdés, A., Sepúlveda, L., De la Cruz, R., Prado-Barragán, A., Aguilar-González, M. A., et al., (2014). Potential use of different agro-

industrial by-products as supports for fungal ellagitannase production under solid-state fermentation. *Food Bioprod. Process*, *92*(4), 376–382.

136. Ascacio-Valdés, J. A., Buenrostro-Figueroa, J. J., Aguilera-Carbo, A., Prado-Barragán, A., Rodríguez-Herrera, R., & Aguilar, C. N., (2011). Ellagitannins: Biosynthesis, biodegradation and biological properties. *J. Med. Plants Res.*, *5*(19), 4696–4703.

137. Aguilera-Carbó, A., Augur, C., Prado-Barragan, L., Aguilar, C., & Favela-Torres, E., (2008). Extraction and analysis of ellagic acid from novel complex sources. *Chem. Pap.*, *62*(4), 440–444.

138. Robledo, A., Aguilera-Carbó, A., Rodriguez, R., Martinez, J. L., Garza, Y., & Aguilar, C. N., (2008). Ellagic acid production by *Aspergillus niger* in solid-state fermentation of pomegranate residues. *J. Ind. Microbiol. Biotechnol.*, *35*(6), 507–513.

139. Koponen, J. M., Happonen, A. M., Mattila, P. H., & Törrönen, A. R., (2007). Contents of anthocyanins and ellagitannins in selected foods consumed in finland. *J. Agric. Food Chem.*, *55*(4), 1612–1619.

140. Seeram, N., Lee, R., Hardy, M., & Heber, D., (2005). Rapid large scale purification of ellagitannins from pomegranate husk, a by-product of the commercial juice industry. *Sep. Purif. Technol.*, *41*(1), 49–55.

141. Gómez-Romero, M., Arráez-Román, D., Segura-Carretero, A., & Fernández-Gutiérrez, A., (2007). Analytical determination of antioxidants in tomato: Typical components of the Mediterranean diet. *J. Sep. Sci.*, *30*(4), 452–461.

142. Shui, G., & Leong, L. P., (2006). Residue from star fruit as valuable source for functional food ingredients and antioxidant nutraceuticals. *Food Chem.*, *97*(2), 277–284.

143. Rice-Evans, C., Miller, N., & Paganga, G., (1997). Antioxidant properties of phenolic compounds. *Trends in Plant Sci.*, *2*(4), 152–159.

144. Tachakittirungrod, S., Okonogi, S., & Chowwanapoonpohn, S., (2007). Study on antioxidant activity of certain plants in Thailand: Mechanism of antioxidant action of guava leaf extract. *Food Chem.*, *103*(2), 381–388.

145. Rosenthal, A., Pyle, D. L., Niranjan, K., Gilmour, S., & Trinca, L., (2001). Combined effect of operational variables and enzyme activity on aqueous enzymatic extraction of oil and protein from soybean. *Enzyme Microb. Technol.*, *28*(6), 499–509.

146. Singh, R., Sarker, B., Kumbhar, B., Agrawal, Y., & Kulshreshtha, M., (1999). Response surface analysis of enzyme assisted oil extraction factors for sesame, groundnut and sunflower seeds. *J. Food Sci. Technol.*, *36*(6), 511–514.

147. Kapasakalidis, P. G., Rastall, R. A., & Gordon, M. H., (2009). Effect of a cellulase treatment on extraction of antioxidant phenols from black currant (*Ribes nigrum* L.) pomace. *J. Agric. Food Chem.*, *57*(10), 4342–4351.

148. Dulf, F. V., Vodnar, D. C., & Socaciu, C., (2016). Effects of solid-state fermentation with two filamentous fungi on the total phenolic contents, flavonoids, antioxidant activities and lipid fractions of plum fruit (*Prunus domestica* L.) by-products. *Food Chem.*, *209*, 27–36.

149. Singh, H. B., Singh, B. N., Singh, S. P., & Nautiyal, C. S., (2010). Solid-state cultivation of *Trichoderma harzianum* NBRI-1055 for modulating natural antioxidants in soybean seed matrix. *Biores. Technol.*, *101*(16), 6444–6453.

150. Zhang, Z., Lei, Z., L, Y., L, Z., & Chen, Y., (2008). Chemical composition and bioactivity changes in stale rice after fermentation with *Cordyceps sinensis*. *J. Biosci. Bioeng.*, *106*(2), 188–193.

151. Poyart-Salmeron, C., Carlier, C., Trieu-Cuot, P., Courvalin, P., & Courtieu, A. L., (1990). Transferable plasmid-mediated antibiotic resistance in *Listeria monocytogenes*. *The Lancet, 335*(8703), 1422–1426.

152. Meng, J., Zhao, S., Doyle, M. P., & Joseph, S. W., (1998). Antibiotic resistance of *Escherichia coli* O157: H7 and O157: NM isolated from animals, food, and humans. *J. Food Prot., 61*(11), 1511–1514.

153. Sobota, A., (1984). Inhibition of bacterial adherence by cranberry juice: Potential use for the treatment of urinary tract infections. *J. Urol., 131*(5), 1013–1016.

154. Acamovic, T., & Stewart, C., (1999). In: *Plant Phenolic Compounds and Gastrointestinal Micro-Organisms* (pp. 127–129). Aciar Proceedings, Aciar.

155. Vattem, D. A., Lin, Y. T., Labbe, R. G., & Shetty, K., (2004). Antimicrobial activity against select food-borne pathogens by phenolic antioxidants enriched in cranberry pomace by solid-state bioprocessing using the food grade fungus *Rhizopus oligosporus*. *Process Biochem., 39*(12), 1939–1946.

156. Akiyama, H., Fujii, K., Yamasaki, O., Oono, T., & Iwatsuki, K., (2001). Antibacterial action of several tannins against *Staphylococcus aureus*. *J. Antimicrob. Chemo., 48*(4), 487–491.

157. Sepúlveda, L., Ascacio, A., Rodríguez-Herrera, R., Aguilera-Carbó, A., & Aguilar, C. N., (2011). Ellagic acid: Biological properties and biotechnological development for production processes. *Afr. J. Biotechnol., 10*(22), 4518–4523.

158. Machado, T. D. B., Leal, I. C. R., Amaral, A. C. F., Santos, K. R. N. D., Silva, M. G. D., & Kuster, R. M., (2002). Antimicrobial ellagitannin of *Punica granatum* fruits. *J. Braz. Chem. Soc., 13*, 606–610.

159. Namiki, M., (1990). Antioxidants/antimutagens in food. *Crit. Rev. Food Sci. Nutr., 29*(4), 273–300.

160. Hochstein, P., & Atallah, A. S., (1988). The nature of oxidants and antioxidant systems in the inhibition of mutation and cancer. *Mutation Research/Fundamental and Molecular Mechanisms of Mutagenesis, 202*(2), 363–375.

161. Ames, B. N., (1985). Dietary carcinogens and anticarcinogens: Oxygen radicals and degenerative diseases. In: Whipple, C., & Covello, V. T., (eds.), *Risk Analysis in the Private Sector* (pp. 297–321). Springer US: Boston, MA.

162. Stavric, B., (1994). Antimutagens and anticarcinogens in foods. *Food Chem. Toxicol., 32*(1), 79–90.

163. Negi, P. S., Jayaprakasha, G. K., & Jena, B. S., (2003). Antioxidant and antimutagenic activities of pomegranate peel extracts. *Food Chem., 80*(3), 393–397.

164. Ayrton, A. D., Lewis, D. F. V., Walker, R., & Ioannides, C., (1992). Antimutagenicity of ellagic acid towards the food mutagen IQ: Investigation into possible mechanisms of action. *Food Chem. Toxicol., 30*(4), 289–295.

165. Bu-Abbas, A., Clifford, M. N., Walker, R., & Ioannides, C., (1994). Marked antimutagenic potential of aqueous green tea extracts: Mechanism of action. *Mutagen., 9*(4), 325–331.

166. Krul, C., Luiten-Schuite, A., Tenfelde, A., Van Ommen, B., Verhagen, H., & Havenaar, R., (2001). Antimutagenic activity of green tea and black tea extracts studied in a dynamic *in vitro* gastrointestinal model. *Mutagen. Fund. Mol. Mech. Mut., 474*(1–2), 71–85.

167. Thériault, M., Caillet, S., Kermasha, S., & Lacroix, M., (2006). Antioxidant, antiradical and antimutagenic activities of phenolic compounds present in maple products. *Food Chem.*, *98*(3), 490–501.

168. Kaur, I. P., & Saini, A., (2000). Sesamol exhibits antimutagenic activity against oxygen species mediated mutagenicity. *Mutat. Res. Gen. Toxicol. Environ. Mutagen.*, *470*(1), 71–76.

169. Valdez-Morales, M., Espinosa-Alonso, L. G., Espinoza-Torres, L. C., Delgado-Vargas, F., & Medina-Godoy, S., (2014). Phenolic content and antioxidant and antimutagenic activities in tomato peel, seeds, and byproducts. *J. Agric. Food Chem.*, *62*(23), 5281–5289.

170. Seeram, N. P., Adams, L. S., Zhang, Y., Lee, R., Sand, D., Scheuller, H. S., et al., (2006). Blackberry, black raspberry, blueberry, cranberry, red raspberry, and strawberry extracts inhibit growth and stimulate apoptosis of human cancer cells *in vitro*. *J. Agric. Food Chem.*, *54*(25), 9329–9339.

171. Silalahi, J., (2002). Anticancer and health protective properties of citrus fruit components. *Asia Pac. J. Clin. Nutr.*, *11*(1), 79–84.

172. Ahn, M. J., Kim, C. Y., Lee, J. S., Kim, T. G., Kim, S. H., Lee, C. K., et al., (2002). Inhibition of HIV-1 integrase by galloyl glucoses from *Terminalia chebula* and flavonol glycoside gallates from *Euphorbia pekinensis*. *Planta Med.*, *68*(5), 457–459.

173. Savi, L. A., Leal, P. C., Vieira, T. O., Rosso, R., Nunes, R. J., Yunes, R. A., et al., (2005). Evaluation of anti-herpetic and antioxidant activities, and cytotoxic and genotoxic effects of synthetic alkyl-esters of gallic acid. *Arzneimittel for Schung.*, *55*(01), 66–75.

174. Serkedjieva, J., (2000). Combined antiinfluenza virus activity of Flos verbasci infusion and amantadine derivatives. *Phytother. Res.*, *14*(7), 571–574.

175. Benavente-García, O., Castillo, J., Marin, F. R., Ortuño, A., Del Río, J. A., (1997). Uses and properties of citrus flavonoids. *J. Agric. Food Chem.*, *45*(12), 4505–4515.

176. Cefarelli, G., D'Abrosca, B., Fiorentino, A., Izzo, A., Mastellone, C., Pacifico, S., et al., (2006). Free-radical-scavenging and antioxidant activities of secondary metabolites from reddened Cv. annurca apple fruits. *J. Agric. Food Chem.*, *54*(3), 803–809.

177. Cos, P., Berghe, D. V., Bruyne, T. D., & Vlietinck, A. J., (2003). Plant substances as antiviral agents: An update (1997–2001). *Curr. Org. Chem.*, *7*(12), 1163–1180.

178. Gonçalves, J. L. S., Leitão, S. G., Monache, F. D., Miranda, M. M. F. S., Santos, M. G. M., Romanos, M. T. V., et al., (2001). *In vitro* antiviral effect of flavonoid-rich extracts of *Vitex polygama* (Verbenaceae) against acyclovir-resistant herpes simplex virus type 1. *Phytomedicine*, *8*(6), 477–480.

179. Okuda, T., Yoshida, T., & Hatano, T., (1992). Polyphenols from Asian plants. In: Huang, M. T., Ho, C. T., & Lee, C. Y., (eds.), *Phenolic Compounds in Food and Their Effects on Health II* (Vol. 507, pp. 160–183). American Chemical Society.

180. Makare, N., Bodhankar, S., & Rangari, V., (2001). Immunomodulatory activity of alcoholic extract of *Mangifera indica* L. in mice. *J. Ethnopharmacol.*, *78*(2), 133–137.

# CHAPTER 2

# ADVANCES ON FERMENTATION PROCESSES FOR THE PRODUCTION OF BIOACTIVE COMPOUNDS IN FOOD BIOTECHNOLOGY

LEONARDO SEPÚLVEDA,[1] JOSÉ JUAN BUENROSTRO-FIGUEROA,[2] JUAN A. ASCACIO-VALDÉS,[1] ANTONIO F. AGUILERA-CARBÓ,[3] and CRISTÓBAL N. AGUILAR[1]

[1] Food Research Department, Autonomous University of Coahuila, Saltillo, Coahuila, México

[2] Research Center for Food and Development A.C., Delicias, Chihuahua, México

[3] Department of Animal Nutrition, Agrarian Autonomous University "Antonio Narro," Saltillo, Coahuila, México

## ABSTRACT

In the food industry, microbial processes are of great relevance because they use low-cost agroindustrial waste for the production of bioactive compounds. Bioactive compounds or secondary metabolites from plant origin are also called phytochemicals, they are compounds with important biological properties and human health benefits such as antimicrobial, antiviral, antioxidant, anticancer, among others. In addition, they are compounds with high added value because they can be applied in different areas of research for example in organic chemistry until emerging technologies. This chapter will explain topics related to the production of bioactive compounds using solid-state fermentation and submerged fermentation. On the other hand, the microorganisms more used in the microbial processes and physicochemical parameters that most influence the optimization in the fermentation processes. Finally, the most recent applications of bioactive compounds obtained by fermentation processes used in food biotechnology.

## 2.1   GENERAL ASPECTS ON THE PRODUCTION OF BIOACTIVE COMPOUNDS BY FERMENTATION PROCESSES

In the last decades the interest has grown to obtain bioactive compounds of high added value, under this context, the fermentation processes in solid and submerged state culture play a primordial role for the production of these compounds using agroindustrial wastes as source of carbon and energy by microorganisms with an enzymatic machinery capable of biodegrading the compounds present in the system. This section mentions some advances related to the production of bioactive compounds using fermentation systems. For example, a *Streptomyces* strain was isolated and identified from the soil. During the solid-state fermentation using this strain were obtained three putative compounds as angumycinones C and D and compound X–14881 E. The best results indicate that they are potentially bioactive against other drug-resistant bacterial pathogens [1]. In another study, the biotreatment of olive pomace in submerged culture with *Streptomyces* sp. S1M3 was evaluated in order to produce lignocellulolytic enzymes and upgrade the nutritional value of olive pomace for incorporating in the livestock feed. The results obtained were that *Streptomyces* sp. S1M3I exhibited activities, of $11.2 \pm 0.12$ U/mL for xylanase, $1.44 \pm 0.02$ U/mL for cellulase and 1.21 9 10–2U/mL for laccase. A significant ($p<0.05$) decrease in the hemicellulose, cellulose, and lignin content was registered. Additionally, an increase in crude protein content with 34.18% and a decrease in total lipid content with 82.23% were registered. The authors concluded that the olive pomace can be valorized by submerged culture of *Actinobacteria* strains; this can be an interesting alternative for biotechnological processes [2]. On the other hand, tangerine residues can be effectively used by converting them into valuable foods/feeds through a procedure of solid-state fermentation. The tangerine residues were evaluated to support the growth of *Lentinus polychrous* Lév. The results indicate that during fermentation, the protein content was significantly increased. The phenolic content and DPPH scavenging activity were gradually reduced. The bioactive compounds of dried and fermented tangerine residues demonstrated a high potential for the development of protein-aceous and antioxidative food/feed ingredients [3]. Further, the production of theabrownins from infusions of sun-dried green tea leaves using a pure culture of *Aspergillus fumigatus* isolated from a solid-state Pu-erh tea fermentation was evaluated. The results showed that theabrownins production in shake-flasks was 158 g/kg sun-dried green tea leaves within 6 days at 45°C in aerobic fermentation. Extracellular polyphenol oxidase and peroxidase of *A. fumigatus* contributed to this bioconversion [4]. In another study, a tannase that was prepared from *Aspergillus niger* by a solid-state fermentation on tea byproducts was investigated to transform the catechins in tea infusion. The tannase could effectively hydrolyze epigallocatechin gallate, gallocatechin gallate, and epicatechin gallate of tea infusion, which showed pH values ranged from 4.8 to 5.3 during the tannase treatment. The results indicated the tannase was effective to transform the catechins. Authors concluded that it is more desirable to follow the dynamic model to control the tannase treatment of tea infusion for a better processing [5]. One of the enzymes related to the hydrolysis of tannins is tannin acyl hydrolase (EC 3.1.1.20), better known as tannase. This enzyme has several applications in the pharmaceutical and food industry that is why studies are carried out for its production using fermentation systems. For example, the production and characterization of a multi-tolerant tannase from *Aspergillus carbonarius* in a fermentation system was evaluated. The best results showed that the filamentous fungus *A. carbonarius* produced high levels of tannase when cultivated under solid-state fermentation using green tea leaves as substrate/carbon source and tap water at a 1:1 ratio as the moisture agent for 72 h at 30°C. They concluded that *A. carbonarius* tannase present properties as multi-tolerance, which highlight its potential for future application [6]. On the other hand, the leaves of *Psidium guajava* L. were co-fermented with *Monascus anka* and *Bacillus* sp. to promote the release of insoluble-bound polyphenol components. The results indicated that total polyphenols and quercetin content were enhanced at the first 8 days of fermentation. This study provided a new way to upgrade polyphenols or other bioactive compounds with higher health beneficial effects [7]. In another study, diosgenin extraction by direct biotransformation with *Penicillium dioscin* was investigated. The best cultivation conditions were: Czapek liquid culture medium without sugar and agar (1,000 ml) + 6.0 g dioscin/6.0 g DL, 30°C, 36 h; solid fermentation of Rhizome of *Dioscorea zingiberensis* mycelia/RDZ of 0.05 g/kg, 30°C, 50 h; the yield of diosgenin was over 90%. According to the results, this biotransformation method was environment-friendly, simple, and energy saving [8]. Wheat grains from different cultivars were fermented with a fungal strain *Aspergillus awamorinakazawa* using solid-state fermentation and evaluated their bioactive properties. The results showed that *A. awamorinakazawa* used for solid-state fermentation was effective for increasing the phenolic content, DPPH radical, total flavonoids content, ABTS activity and metal chelating activities of wheat cultivars (both *Triticum* spp.). For this reason, the fermentation can be used as a tool to develop wheat as health food ingredient that possess multifunctional properties [9]. Finally, a solid-state fermentation of chickpeas with *Cordyceps militaris* SN–18 was performed for the first time. The effects of fermentation on chickpeas were investigated in terms of total phenolic and saponin contents, antioxidant activities, and DNA damage protection through various solvent extracts. The results showed compared to the unfermented samples, fermented chickpea extracts had higher total phenolic and saponin contents, and greater antioxidant and DNA damage protective activities. They concluded that the fermentation process with *Cordyceps militaris* enhanced antioxidant capacities of chickpeas and thus was considered of great potential for the food industry [10].

The solid and submerged fermentation processes are a powerful tool in the area of food and biotechnology to take advantage of the compounds present in the substrate with the help of a microorganism capable of biodegrading these compounds and releasing bioactive compounds of high added value with important applications in pharmacy, medicine, and food.

## 2.1.2 IMPORTANCE OF THE MICROORGANISM IN THE FERMENTATION PROCESSES

One of the essential stages of fermentation processes is the proper selection of a microorganism that is capable of adapting, growing, and releasing compounds present in the substrate. In this section we will focus only on the fermentation processes where fungal strains are used, since they are one of the most used microorganisms for the production of bioactive compounds. Table 2.1 shown some fungi used in different fermentation systems for the production of some bioactive compound.

**TABLE 2.1** Examples of Some Investigations on Use of Fungi to Produce Bioactive Compounds From Industrial Application. Fungi Produces Bioactive Compounds or Release Them?

| Fungi | Fermentation type | Conditions | Bioactive compound | Reference |
|---|---|---|---|---|
| Aspergillus fumigattus AFGRD105 | Submerged | Temperature 45°C, sodium dihydrogen phosphate 14 g, moisture 25% | Melanin | Raman et al. [11] |
| Fusarium solani pisi | Solid-state | pH 8, temperature 50°C, particle size 0.8–1.25 mm, solid-to-crude enzymes solution ratio 1:30. | Lycopene | Azabou et al. [12] |
| Aspergillus heteromorphus MTCC 8818 | Solid-state | Czapek-Dox medium and pH 5.5, relativity humidity 70%, 30°C and 96 h | Tannase | Beniwal et al. [13] |
| Xylaria nigripes | Solid-state | Inoculum 10%, volume 1L, temperature 25°C, humidity 60%, 30 days | Phenols and flavonoids | Divate et al. [14] |
| Aspergillus niger GH1 | Solid and submerged state | Polyurethane foam as support, temperature 30°C, 72 h (solid fermentation). Volume 30 mL, temperature 30°C, 72 h at 250 rpm (submerged fermentation) | Phenolics | Chávez-González et al. [15] |
| Antrodia cinnamomea | Solid-state | Inoculum 10%, temperature 25°C, buckwheat powder 28 g and 20 mL of water | Polysaccharides intracellular, polyphenols, and triterpenoids | Yang et al. [16] |
| Rhizopus oryzae | Solid-state | Temperature 30°C, incubation area 2340 cm³, substrate density 0.1 dry fruit g/ cm², inoculum 1%, 72 h | Proteins | Ibarruri and Hernández [17] |
| Paecilomyces variotii | Solid-state | Particle size 1.86 mm, citrus residues 10 g and 10 mL of water, inoculum 9 x 10⁶ spores, temperature 30°C, relative humidity 90% | Phenolics | Madeira Jr et al. [18] |

**TABLE 2.1** (Continued)

| Fungi | Fermentation type | Conditions | Bioactive compound | Reference |
|---|---|---|---|---|
| *Neurospora sitophila* | Submerged and solid-state | Steam exploded wheat straw 5 g and 15 mL Vogel's medium, 5 g of sucrose, 1 mL spore suspension, temperature 30°C (solid fermentation). Volume 100 mL, spore solution 1 mL, 180 rpm and temperature 30°C (submerged fermentation) | Proteins | Li et al. [19] |
| *Monascus ruber* CCT 3802 | Submerged | Volume 3.6 L, inoculum 0.4 L (10%), temperature 30°C, 300 rpm, aeration 0.6 vvm (volume of air per volume of batch per minute), pH initial 3.0 and 5.0 | Natural pigments | Vendruscolo et al. [20] |

Filamentous fungi are the most commonly used microorganisms for the production of bioactive compounds with high added value. For example, the effect of different sources of ellagitannins (cranberry, creosote bush, and pomegranate) on the efficiency of ellagic acid release by different *Aspergillus niger* strains (GH1, PSH, and HT4) were analyzed. Polyurethane foam was used as support for solid-state culture with column reactors. When pomegranate polyphenols were used, a maximum value of ellagic acid (350.21 mg/g) was reached with *A. niger* HT4 in solid-state culture. This results showed that the best source for releasing ellagic acid were pomegranate polyphenols and *A. niger* HT4 strain, due to the ability to degrade these compounds for obtaining a potent bioactive molecule such as ellagic acid [21]. In another study, fungi isolated from less explored forest soil ecosystem of Northeast India were studied for the production of antimicrobial metabolites. Out of the 68 fungi isolated from forest soil of Manipur, among them, *Aspergillus terreus* (IBSD-F4) showed the most significant activity against *Staphylococcus aureus* (ATCC–25923), *Bacillus anthracis* (IBSDC370), *Pseudomonas fluorescens* (ATCC–13525), *Salmonella typhimurium* (ATCC–14028), *Escherichia coli* (ATCC–25922), and *Candida albicans* (ATCC–10231). The active metabolite obtained from submerged fermentation system was identified as *Sclerotionigrin A*. The authors mention on the importance of exploring microbes from forest soil for identification of bioactive metabolites for future drug development [22]. By last, the feeds produced by solid-state fermentation of the industrial solid wastes (fruits of milk thistle *Silybum marianum*) was evaluated. The protein content of the fermented feed from a combination of *Aspergillus niger* and *Candida tropicalis* was the highest among the examined strains. The optimal process parameters for protein using *A. niger* and *C. tropicalis* were incubation temperature of 30.8°C, fermentation time of 87 h, and initial moisture content of 59.7%. They concluded that the improvement was achieved through the combination of strains by solid-state fermentation from solid wastes [23].

## 2.1.3  FACTORS THAT INFLUENCE IN THE OPTIMIZATION ON THE FERMENTATION PROCESSES

One of the critical stages in fermentation processes is the evaluation of the physicochemical factors and how they influence the release or accumulation of a bioactive compound. This section mentions some examples of research on the optimization and effect of temperature, pH, and source carbon, among others factors in solid and submerged fermentation for the production of a bioactive compound with application in the food or pharmaceutical industry. The ellagic acid accumulation in solid-state culture using *Aspergillus niger* GH1 and powdered pomegranate peel as a support was evaluated. Various culture conditions (temperature, initial moisture, levels of inoculum and concentration of salts) were evaluated using a Plackett–Burman design followed by a Central Composite Design for enhancing the ellagic acid accumulation. Temperature, $MgSO_4$ and KCl concentration were identified as significant parameters on ellagic acid accumulation. The effect of temperature in the process can be helping activate some enzymes related with biodegradation of polyphenols. The salts in the medium can act like nutrients for optimal growth of the fungi [24]. Temperature is one of the most important process variables affecting solid-state fermentation, because microbial growth under aerobic conditions results in the release of metabolic heat and the control of temperature in the laboratory is a great problem. At high levels, this can cause denaturation of the enzymes produced, as well as other effects of microorganism growth and metabolite production [25]. The addition of trace metal ions and vitamins in the fermentation medium was reported to be beneficial for fungal growth and metabolism. Metal ions such as Mg, $Zn^{2+}$, $Fe^{2+}$, $Cu^{2+}$, and $Mn^{2+}$; and vitamins such as biotin and riboflavin are used as cofactors or activators for some enzymes involved in catabolic and biosynthetic pathways such as the TCA cycle [26]. In other research, the optimal fermentation conditions and medium for the production of bioactive polysaccharides from the mycelium of *Cordyceps sinensis* fungus UM01 and their purification were investigated by an orthogonal design. The best results showed that optimal temperature, initial pH, rotation speed, medium capacity (ratio of medium volume to the volume of reactor) and inoculums volume for mycelium growth were 15°C, pH 6.0, 150 rpm, 2/5 (v/v), and 3% (v/v), respectively. One of the determining factors in the process is pH control, the author's mention that, the effect of this factor in the process is very important because the medium pH may affect cell membrane function, cell morphology and structure, the uptake of various nutrients and product of biosynthesis [27]. The solid-state fermentation parameters of defatted soybean flour with *Aspergillus oryzae* IOC 3999/1998 or *Monascus purpureus* NRRL 1992 was evaluated using a rotational central composite experimental design to optimize the production of β-glucosidase. Factors evaluated were initial pH of defatted soybean flour, the volume of water added to 10 g of defatted soybean flour and incubation temperature. The highest production of β-glucosidase for both strains occurred when adding 10 mL of water to the defatted soybean flour, incubating at 30°C and using 6.0 as the initial defatted soybean flour pH. One of the essential factors in the fermentation process was the water content. One of the theories that affects during fermentation is due to high moisture content decreases the porosity and diffusion of oxygen, while the low content may negatively influence microorganism growth because of lower nutrient solubility and rapid water loss by evaporation [28]. In another study, the medium optimization was performed to enhance production of citric acid through the selecting of five strains of *Aspergillus niger* using hydrolyzed cassava peel medium. The physicochemical parameters for citric acid production were optimized using a Central Composite Design. The experimental citric acid yields were greatly influenced by fermentation time, pH, inoculum size, and substrate concentration. The percentage of substrate in the fermentation process is very significant because this factor is the source carbon and energy of microorganism to produce citric acid, in this study use cassava peels and contains high-level of carbohydrates [29]. Effects of various culture conditions on lovastatin production were investigated by *Aspergillus terreus* (KM017963) grown under solid-state fermentation with wheat bran. Lovastatin production was influenced by various physical factors such as pH, temperature, and nutritional factors such as carbon, nitrogen, metal ions/salts, etc. The best condition in this study was initial pH of 6.0, growth temperature of 28–30°C, inoculum size of $1 \times 10^8$ spores/mL as the optimal physiological culture conditions for maximal production of lovastatin by *A. terreus* (KM017963). One the effect significant in the process was inoculum. Higher inoculum volume ($1 \times 10^{10}$ spores/mL) leads to decreased lovastatin production when compared to $1 \times 10^8$ spores/mL [30].

The optimization of the factors in a fermentation system is a topic very important to explain how they affect the production of bioactive compounds. Besides, each study is particularly original, this means that there may be multiple theories to explain the effect of the factors in the fermentation process.

## 2.1.4  APPLICATIONS OF BIOACTIVE COMPOUNDS OBTAINED BY FERMENTATION PROCESSES USED IN FOOD BIOTECHNOLOGY

Strategies have been used to replace the chemical extraction of bioactive compounds of interest in the food area such as polyphenols, and bioprocesses offer an alternative where microorganisms, mainly filamentous fungi, are involved. The bioprocess that is carried out causes changes in the bioactive compounds present in different plant sources, which allows a better bioavailability and facilitates the recovery of bioactive compounds.

There is scarce information on the use of bioprocesses to obtain bioactive compounds, several authors have mentioned in their works that it is crucial to carry out studies that provide information [31, 32]. However, diverse studies describing the use and bioprocesses have been reported, which refer to the production of these compounds by the action of enzymes produced by fungi. In this section, we will talk about some applications and perspectives of application of some bioactive compounds obtained through bioprocesses.

In general, bioactive compounds produced by bioprocesses, such as ellagic acid, act as antioxidants by donating electrons that free radicals receive, this prevents the reaction between free radicals and human cells [33, 34]. This represents a very important application in the food area because compounds such as ellagic acid represent an option for the generation of functional foods.

Another application of bioactive compounds obtained by bioprocesses is as anticancer compounds, because they react against known carcinogenic compounds, such as benzo-[a]-pyrene-7,8-diol-9,10-epoxide (BPDE), inhibiting the mutation of healthy cells, initial step in cancer development [35]. It has been demonstrated that these compounds decrease the multiplicity of colon tumors in mice and the incidence of carcinoma in the mouse bladder when administered orally in the diet [36]. In addition, bioactive compounds have been reported to be capable of the inhibition of malignant cells proliferation from skin cancer (melanoma), breast, stomach, lung, esophagus, liver, and other cancers [37–40].

The application as anticancer of the bioactive compounds is not the only one that has been reported. Application as antiviral potentials has also been reported; compounds such as ellagic acid have been reported as antiviral against the herpes simplex virus (anti-HSV). Studies have been conducted on the antiviral activity of bioactive compounds obtained from pomegranate, such as ellagic acid, caffeic acid, luteolin, and punicalagin, which block the replication of human influenza A virus. Other authors have concluded that these compounds are involved in the viruses' inhibition not only in vitro, in vivo also, preventing viral infections [41].

One of the most important applications of these compounds which are produced using bioprocess is the application as antimicrobial agents, particularly against foodborne pathogenic microorganisms. It has been reported that compounds derived from ellagic acid have activity against certain pathogenic bacteria, such as *Klebsiella pneumoniae, Bacillus cereus, Escherichia coli, Salmonella typhi* and *Salmonella pyogenes* [42]. Also, these compounds showed the ability to inhibit the growth of *Pseudomonas aeruginosa* and *Bacillus subtilis*. Another example is punicalagin, a precursor of ellagic acid, which showed a clear zone of inhibition against methicillin-resistant strains of *Staphylococcus aureus*. Punicalagin, punicalin, gallagic acid, and ellagic acid, which showed antimicrobial activity against bacteria such as *Escherichia coli, Candida albicans* and *Crypotococcus neoformans* and the fungus *Aspergillus fumigatus* were obtained by fermentation of *Punica granatum* L. [38, 43].

An interesting application of this type of compounds is their use in the formulation of edible coverings to improve the shelf life of some foodstuffs. For example, ellagic acid that is produced by bioprocesses has been added to edible coverings for application in avocados, where a decrease in changes in appearance, solids content, pH, $a_w$, luminosity, and weight loss was observed, maintaining the quality. Thus, it was demonstrated that this application can work to prolong the shelf life of the analyzed avocados [44].

Finally, phenolic compounds are intracellular metabolites that interact with the cell wall components (cellulose, hemicellulose, and pectins), often linked to them, which prevents their easy release. During solid-state fermentation processes, microorganisms are able to produce different enzymes, such as cellulase, β-glucosidase, xylanase, pectinase, tannase, and ellagitannase, which in a synergistic activity help to break and degrade the cell wall structure, releasing the linked compounds [45, 46]. Research has been focused as tools to optimize the extraction of phenolic compounds from plants and agroindustrial byproducts [47–51].

In conclusion, it is clear that the applications of the bioactive compounds obtained by bioprocesses can be very broad and diverse, it is important to focus the scientific work in the development of bioprocesses to obtain the compounds and develop the strategies to evaluate the applications, such as nutraceuticals, therapeutic agents added to foods, preventive agents of cellular diseases by the formulation of functional foods.

## KEYWORDS

- bioactive compounds
- ellagic acid
- phenolic compounds

## REFERENCES

1. Su, H., Shao, H., Zhang, K., & Li, G., (2016). Antibacterial metabolites from the *Actinomycete streptomyces* sp. P294. *Journal of Microbiology, 54*(2), 131–135.
2. Medouni-Haroune, L., Zaidi, F., Medouni-Adrar, S., Kernou, O. N., Azzouz, S., & Kecha, M., (2017). Bioconversion of olive pomace by submerged cultivation of streptomyces sp. S1M3I. *Proceedings of the National Academy of Sciences, India Section B: Biological Sciences,* 1–9.
3. Nitayapat, N., Prakarnsombut, N., Lee, S. J., & Boonsupthip, W., (2015). Bioconversion of tangerine residues by solid-state fermentation with Lentinus polychrous and drying the final products. *LWT – Food Science and Technology, 63*(1), 773–779.
4. Wang, Q., Gong, J., Chisti, Y., & Sirisansaneeyakul, S., (2014). Bioconversion of tea polyphenols to bioactive theabrownins by *Aspergillus fumigatus*. *Biotechnology Letters, 36*(12), 2515–2522.
5. Ni, H., Chen, F., Jiang, Z. D., Cai, M. Y., Yang, Y. F., Xiao, A. F., et al., (2015). Biotransformation of tea catechins using *Aspergillus niger tannase* prepared by solid-state fermentation on tea byproduct. *LWT – Food Science and Technology, 60*(2), 1206–1213.
6. Valera, L. S., Jorge, J. A., & Guimarães, L. H. S., (2015). Characterization of a multi-tolerant tannin acyl hydrolase II from *Aspergillus carbonarius* produced under solid-state fermentation. *Electronic Journal of Biotechnology, 18*(6), 464–470.
7. Wang, L., Bei, Q., Wu, Y., Liao, W., & Wu, Z., (2017). Characterization of soluble and insoluble-bound polyphenols from *Psidium guajava* L. leaves co-fermented with *Monascus anka* and *Bacillus* sp. and their bio-activities. *Journal of Functional Foods, 32*, 149–159.

8. Dong, J., Lei, C., Lu, D., & Wang, Y., (2015). Direct biotransformation of dioscin into diosgenin in rhizome of dioscorea zingiberensis by *penicillium dioscin*. *Indian Journal of Microbiology, 55*(2), 200–206.

9. Sandhu, K. S., Punia, S., & Kaur, M., (2016). Effect of duration of solid-state fermentation by *Aspergillus awamorinakazawa* on antioxidant properties of wheat cultivars. *LWT – Food Science and Technology, 71*, 323–328.

10. Xiao, Y., Xing, G., Rui, X., Li, W., Chen, X., Jiang, M., et al., (2014). Enhancement of the antioxidant capacity of chickpeas by solid-state fermentation with *Cordyceps militaris* SN-18. *Journal of Functional Foods, 10*, 210–222.

11. Raman, N. M., Shah, P. H., Mohan, M., & Ramasamy, S., (2015). Improved production of melanin from *Aspergillus fumigatus* AFGRD105 by optimization of media factors. *AMB Express, 5*(1), 72.

12. Azabou, S., Abid, Y., Sebii, H., Felfoul, I., Gargouri, A., & Attia, H., (2016). Potential of the solid-state fermentation of tomato by products by Fusarium solani pisi for enzymatic extraction of lycopene. *LWT – Food Science and Technology, 68*, 280–287.

13. Beniwal, V., Rajesh, G. G., Kumar, A., & Chhokar, V., (2013). Production of tannase through solid-state fermentation using Indian rosewood (*Dalbergia sissoo*) sawdust— a timber industry waste. *Annals of Microbiology, 63*(2), 583–590.

14. Divate, R. D., Wang, C. C., Chou, S. T., Chang, C. T., Wang, P. M., & Chung, Y. C., (2017). Production of *Xylaria nigripes*-fermented grains by solid-state fermentation and an assessment of their resulting bioactivity. *LWT – Food Science and Technology, 81*, 18–25.

15. Chávez-González, M. L., Guyot, S., Rodríguez-Herrera, R., Prado-Barragán, A., & Aguilar, C. N., (2014). Production profiles of phenolics from fungal tannic acid biodegradation in submerged and solid-state fermentation. *Process Biochemistry, 49*(4), 541–546.

16. Yang, F. C., Yang, Y. H., & Lu, H. C., (2013). Enhanced antioxidant and antitumor activities of cultured *Antrodia cinnamomea* with cereal substrates in solid-state fermentation. *Biochemical Engineering Journal, 78*, 108–113.

17. Ibarruri, J., & Hernández, I., (2017). *Rhizopus oryzae* as fermentation agent in food derived sub-products. *Waste and Biomass Valorization,* 1–9.

18. Madeira, J. V., Nakajima, V. M., Macedo, J. A., & Macedo, G. A., (2014). Rich bioactive phenolic extract production by microbial biotransformation of Brazilian citrus residues. *Chemical Engineering Research and Design, 92*(10), 1802–1810.

19. Li, Y., Peng, X., & Chen, H., (2013). Comparative characterization of proteins secreted by *Neurospora sitophila* in solid-state and submerged fermentation. *Journal of Bioscience and Bioengineering, 116*(4), 493–498.

20. Vendruscolo, F., Luise Müller, B., Esteves Moritz, D., De Oliveira, D., Schmidell, W., & Luiz Ninow, J., (2013). Thermal stability of natural pigments produced by *Monascus ruber* in submerged fermentation. *Biocatalysis and Agricultural Biotechnology, 2*(3), 278–284.

21. Sepúlveda, L., De la Cruz, R., Buenrostro, J. J., Ascacio-Valdés, J. A., Aguilera-Carbó, A. F., Prado, A., et al., (2016). Effect of different polyphenol sources on the efficiency of ellagic acid release by *Aspergillus niger*. *Revista Argentina de Microbiología, 48*(1), 71–77.

22. Devi, S. I., Lotjem, H., Devi, E. J., Potshangbam, M., Ngashangva, N., Bora, J., et al., (2017). Bio-mining the forest ecosystem of North East India for identification of antimicrobial metabolites from fungi through submerged fermentation. *Bioresource Technology, 241,* 1168–1172.

23. Li, F., Li, F., Zhao, T., Mao, G., Zou, Y., Zheng, D., et al., (2013). Solid-state fermentation of industrial solid wastes from the fruits of milk thistle *Silybum marianum* for feed quality improvement. *Applied Microbiology and Biotechnology, 97*(15), 6725–6737.

24. Sepúlveda, L., Aguilera-Carbó, A., Ascacio-Valdés, J. A., Rodríguez-Herrera, R., Martínez-Hernández, J. L., & Aguilar, C. N., (2012). Optimization of ellagic acid accumulation by *Aspergillus niger* GH1 in solid-state culture using pomegranate shell powder as a support. *Process Biochemistry, 47*(12), 2199–2203.

25. Farinas, C. S., (2015). Developments in solid-state fermentation for the production of biomass-degrading enzymes for the bioenergy sector. *Renewable and Sustainable Energy Reviews, 52,* 179–188.

26. Mondala, A. H., (2015). Direct fungal fermentation of lignocellulosic biomass into itaconic, fumaric, and malic acids: Current and future prospects. *Journal of Industrial Microbiology & Biotechnology, 42*(4), 487–506.

27. Wang, L. Y., Cheong, K. L., Wu, D. T., Meng, L. Z., Zhao, J., & Li, S. P., (2015). Fermentation optimization for the production of bioactive polysaccharides from *Cordyceps sinensis* fungus UM01. *International Journal of Biological Macromolecules, 79,* 180–185.

28. Handa, C. L., Couto, U. R., Vicensoti, A. H., Georgetti, S. R., & Ida, E. I., (2014). Optimisation of soy flour fermentation parameters to produce β-glucosidase for bioconversion into aglycones. *Food Chemistry, 152,* 56–65.

29. Adeoye, A. O., Lateef, A., & Gueguim-Kana, E. B., (2015). Optimization of citric acid production using a mutant strain of *Aspergillus niger* on cassava peel substrate. *Biocatalysis and Agricultural Biotechnology, 4*(4), 568–574.

30. Kamath, P. V., Dwarakanath, B. S., Chaudhary, A., & Janakiraman, S., (2015). Optimization of culture conditions for maximal lovastatin production by *Aspergillus terreus* (KM017963) under solid-state fermentation. *HAYATI Journal of Biosciences, 22*(4), 174–180.

31. Scalbert, A., (1992). Quantitative methods for the estimation of tannins in plant tissues. In: Hemingway, R. W., & Laks, P. E., (eds.), *Plant Polyphenols: Synthesis, Properties, Significance* (pp. 259–280). Springer US: Boston, MA.

32. Vivas, N., Laguerre, M., Pianet de Boissel, I., Vivas de Gaulejac, N., & Nonier, M. F., (2004). Conformational interpretation of vescalagin and castalagin physicochemical properties. *Journal of Agricultural and Food Chemistry, 52*(7), 2073–2078.

33. Priyadarsini, K. I., Khopde, S. M., Kumar, S. S., & Mohan, H., (2002). Free radical studies of ellagic acid, a natural phenolic antioxidant. *Journal of Agricultural and Food Chemistry, 50*(7), 2200–2206.

34. Amakura, Y., Okada, M., Tsuji, S., & Tonogai, Y., (2000). Determination of phenolic acids in fruit juices by isocratic column liquid chromatography. *Journal of Chromatography A, 891*(1), 183–188.

35. Hope, S. S., Tate, P. L., Huang, G., Magee, J. B., Meepagala, K. M., Wedge, D. E., et al., (2004). Antimutagenic activity of berry extracts. *Journal of Medicinal Food, 7*(4), 450–455.

36. Kelloff, G. J., Boone, C. W., Crowell, J. A., Steele, V. E., Lubet, R., & Sigman, C. C., (1994). Chemopreventive drug development: Perspectives and progress. *Cancer Epidemiology Biomarkers, Prevention, 3*(1), 85.

37. Clifford, M. N., & Scalbert, A., (2000). Ellagitannins – nature, occurrence and dietary burden. *Journal of the Science of Food and Agriculture, 80*(7), 1118–1125.

38. Seeram, N. P., Adams, L. S., Zhang, Y., Lee, R., Sand, D., Scheuller, H. S., et al., (2006). Blackberry, black raspberry, blueberry, cranberry, red raspberry, and strawberry extracts inhibit growth and stimulate apoptosis of human cancer cells in vitro. *Journal of Agricultural and Food Chemistry, 54*(25), 9329–9339.

39. Kashiwada, Y., Nonaka, G. I., Nishioka, I., Chang, J. J., & Lee, K. H., (1992). Antitumor agents, tannins and related compounds as selective cytotoxic agents. *Journal of Natural Products, 55*(8), 1033–1043.

40. Kaur, G., Jabbar, Z., Athar, M., & Alam, M. S., (2006). Punica granatum (pomegranate) flower extract possesses potent antioxidant activity and abrogates Fe-NTA induced hepatotoxicity in mice. *Food and Chemical Toxicology, 44*(7), 984–993.

41. Ruibal, B. I. J., Dubed, E. M., Martínez, L. F., Noa, R. E., Vargas, G. L. M., & Santana, R. J. L., (2003). Inhibición de la replicación del virus de inmunodeficiencia humana por extractos de taninos de Pinus caribaea Morelet. *Revista Cubana de Farmacia, 37*.

42. Atta Ur, R., Ngounou, F. N., Choudhary, M. I., Malik, S., Makhmoor, T., Nur-E-Alam, M., et al., (2001). New antioxidant and antimicrobial ellagic acid derivatives from pteleopsis hylodendron. *Planta Med., 67*(04), 335–339.

43. Nascimento, G. G. F., Locatelli, J., Freitas, P. C., & Silva, G. L., (2000). Antibacterial activity of plant extracts and phytochemicals on antibiotic-resistant bacteria. *Brazilian Journal of Microbiology, 31*, 247–256.

44. Saucedo-Pompa, S., Rojas-Molina, R., Aguilera-Carbó, A. F., Saenz-Galindo, A., Garza, H. D. L., Jasso-Cantú, D., et al., (2009). Edible film based on candelilla wax to improve the shelf life and quality of avocado. *Food Research International, 42*(4), 511–515.

45. Joana Gil-Chávez, G., Villa, J. A., Fernando Ayala-Zavala, J., Basilio Heredia, J., Sepulveda, D., Yahia, E. M., et al., (2013). Technologies for extraction and production of bioactive compounds to be used as nutraceuticals and food ingredients: An overview. *Comprehensive Reviews in Food Science and Food Safety, 12*(1), 5–23.

46. Buenrostro-Figueroa, J. J., Velázquez, M., Flores-Ortega, O., Ascacio-Valdés, J. A., Huerta-Ochoa, S., Aguilar, C. N., et al., (2017). Solid-state fermentation of fig (*Ficus carica* L.) by-products using fungi to obtain phenolic compounds with antioxidant activity and qualitative evaluation of phenolics obtained. *Process Biochemistry, 62*, 16–23.

47. Gómez-García, R., Martínez-Ávila, G. C. G., & Aguilar, C. N., (2012). Enzyme-assisted extraction of antioxidative phenolics from grape (*Vitis vinifera* L.) residues. *3 Biotech., 2*(4), 297–300.

48. Chen, S., Xing, X. H., Huang, J. J., & Xu, M. S., (2011). Enzyme-assisted extraction of flavonoids from *Ginkgo biloba* leaves: Improvement effect of flavonol transglyco-

sylation catalyzed by *Penicillium decumben*s cellulase. *Enzyme and Microbial Technology*, *48*(1), 100–105.

49. Basegmez, H. I. O., Povilaitis, D., Kitrytė, V., Kraujalienė, V., Šulniūtė, V., Alasalvar, C., et al., (2017). Biorefining of blackcurrant pomace into high value functional ingredients using supercritical $CO_2$, pressurized liquid and enzyme assisted extractions. *The Journal of Supercritical Fluids*, *124*, 10–19.

50. Maier, T., Göppert, A., Kammerer, D. R., Schieber, A., & Carle, R., (2008). Optimization of a process for enzyme-assisted pigment extraction from grape (*Vitis vinifera* L.) pomace. *European Food Research and Technology*, *227*(1), 267–275.

51. Zuorro, A., Fidaleo, M., & Lavecchia, R., (2011). Enzyme-assisted extraction of lycopene from tomato processing waste. *Enzyme and Microbial Technology*, *49*(6–7), 567–573.

# CHAPTER 3

# PRODUCTION OF β-GLUCOSIDASE IN SOLID STATE CULTURES AND ITS APPLICATION IN THE FOOD INDUSTRY

MARISOL CRUZ-REQUENA,[1] ANTONIO F. AGUILERA-CARBÓ,[2] MIGUEL MELLADO,[2] EDUARDO GARCÍA-MARTÍNEZ,[2] REYNALDO DE LA CRUZ,[3] and MIGUEL A. MEDINA-MORALES[2]

[1] Research Center, Identification, Guard and Microbiological Analysis, Rio de Janeiro 590, Colonia Latinoamericana, CP 25270, Saltillo Coahuila, México

[2] Animal Nutrition Department, Animal Science Division, Antonio Narro Agarian Autonomous University, CP 25315 Saltillo Coahuila, México, E-mail: miguelmem84@gmail.com

[3] School of Engineering and Sciences, Monterrey Institute of Technology, Ave. Eugenio Garza-Sada 2501, CP 64849, Monterrey, Nuevo León, México

## ABSTRACT

Enzymes are a key factor in food biotechnology and it is a current topic of interest among researchers. Several enzymes are of great importance in food processes, where in this case, β-glucosidase has many uses in this area. For its production, fungi are often utilized; as cellulase producers, where *Trichoderma* and *Aspergillus* genus stand out among producers. The systems for fungal cellulase production are more common in liquid or submerged state cultures, but interest in solid-state processes is increasing. Solid-state cultures have advantages such as less residual water, higher substrate levels, and less energy input. Most of the processes reported for β-glucosidase production use lignocellulosic residues or highly-available plant resources which also represents an advantage. As it is well known, this enzyme has the role of releasing glucose as the final step in cellulosic enzymatic degradation, but it can also hydrolyze other types of compounds, such as phenolic glycosides, glycoproteins, to name a few. Also, its trans-glycosylation activity is of interest due to the ability to synthesize glyco-sylated compounds useful in the food industry. In this chapter, enzymatic reactions of importance in winemaking, flavor improvement, animal feed, among other aspects, will be addressed giving great importance of food applications concerning biotechnology.

## 3.1   INTRODUCTION

There is a high demand for natural and safe compounds oriented at food industry [1]. The use of enzymes in food processing is a very reliable alter-native due to the low repercussions on the environment and low toxicity in byproducts [2, 3]. Many of the food products that are being introduced in the market, have a tendency to acquire a functional food label or a more simplistic approach of an enzyme-processed product [4]. Biotechnological processes are being developed for processing food or to add high-value compounds to food composition that leads to a functional food categoriza-tion of the end-product [5]. To achieve the purpose of enzyme or compounds production, fungi are often used in the bioprocesses required for these objectives [6]. Fungi are versatile microorganisms due to the wide variety of metabolites useful in the food industry. Ample research has been devel-oped using fungi for metabolite production such as cellulolytic enzymes, polyphenol degrading enzymes and phenolic compound production [7]. Many of the bioprocesses involving said products, appeal to implement solid-state cultures [8]. This type of bioprocess is reported to successfully adapt fungi to substrates with low water content, but with enough mois-ture for the microorganism to grow and degrade the substrate employed [9]. Regarding substrates, agricultural residues can be employed as source of nutrients and support for the fungi to grow on or inert materials which contain a carbon source and enough moisture to keep the solid-state char-acteristics [8, 10]. Fungal strains such as *Aspergillus* and *Penicillium* are reported as cellulase producers, from which stands out the β-glucosidase enzyme [11, 12]. As previously said, among cellulases, the most important enzyme is β-glucosidase (EC 3.2.1.21) for its action of releasing glucose in cellulose degradation, reaching the end of the process [13]. There is a versatility of β-glucosidase: it can also degrade glucosylated compounds with relevance in food processing for organoleptic quality increase [14]. Also, this enzyme is reported to manifest transglycosylation activity [15]. In this case, disaccharides or oligosaccharides can be synthesized by transferring glucose monomers to form larger molecules [16]. Another action of the transglycosylation activity is the formation of glucosides that can be of industrial interest such as glucosylation of phenolic compounds [17]. This type of compounds has attributes such as antioxidant activity, among other bioactivities [18]. The objective of this chapter is to discuss production of β-glucosidase by solid-state cultures and its hydrolytic and transglucolytic activities and its use in food industry.

## 3.2  SOLID-STATE FERMENTATION

This bioprocess is defined as a microbial culture where a microorganism grows in a fermentation system in absence of free water or with very low water content. Solid systems are special for microorganisms that grow in environments with high oxygen presence such as the habitat where they were isolated from. In these conditions, higher yields of microbial products are promoted compared to liquid or submerged fermentations [19].

As previously mentioned, suitable microorganisms are required for the development of a solid-state bioprocess. Most of the microbial cultures made in solid-state are carried out using filamentous fungi [10]. Other microorganisms that are able to proliferate in these conditions are yeasts and bacteria such as the *Bacillus* genus. These microbes require a capacity of growth in a media with low water content and/or low water activity which is another aspect taken into account for solid bioprocesses. Being filamentous fungi the most used microorganisms, the genus *Aspergillus*, *Trichoderma*, *Penicillium*, among others; stand out in the production of enzymes or metabolites of interest in the food industry [20–22]. Concerning β-glucosidase production, as it is associated with endoglucanase and cellobiohydrolase, it is expected that β-glucosidase production improves as efforts have been made to enhance cellulase production [23, 24]. *Trichoderma* produces all of the cellulases, but the enzyme of interest is not produced in sufficient amounts to fully degrade the substrate, so *Aspergillus* is used as it can produce β-glucosidase in higher levels [25].

One of the interesting characteristics of solid-state cultures is the possibility of using residual biomass or agroindustry wastes. In these materials, high concentrations of several compounds can be found such as polyphenols, proteins, and polysaccharides, prominently lignocellulose [5, 26]. Agro-residues, which most of the time are underused or just discarded, are considered an ever-increasing pollution problem. Biotechnologically, purpose can be given to these materials as a substrate for microbes and can help mitigate the pollution problem to some extent. Examples of agroindustry residues or biomass used in solid-state bioprocesses are corn stalks, corn cobs, coffee husks, sugarcane bagasse, mango peels, pomegranate husks, nut shells, agave leaf fibers, tarbush leaves, creosote bush leaves, citrus peels, soybean bran; just to mention a few [27–30].

Solid bioprocesses have a wide array of applications in pharmaceutical, feed, and food industry. Solid-state fermentation has been successfully applied in food processing and also its products can be used as additives or to give functional properties [31–33]. This practice has centuries of being incorporated in food preparation. Food ingredients such as beans have been fermented to elaborate typical dishes and traditional food in many parts of the world [6]. Enzyme production is one of the most important applications of solid-state fermentation. Many enzymes have been reported by solid bioprocesses with many applications [34]. A current topic where enzyme has great impact is plant biomass degradation [35], including cellulosic biofuels and high added-value compounds production with a lesser-known application in animal feed processing [31]. In the case of animal feed processing, enzyme production is closely associated to it. The main objective of fermenting raw material for feeding purposes is protein content enrichment and fiber degradability for improving livestock nutritional value of feed [36]. Regarding enzyme production, cellulolytic enzyme production plays a significant role in increasing fiber susceptibility to degradation by mainly producing cellulases and xylanases; among other enzymes [37].

## 3.3  β-GLUCOSIDASE

As it is widely known, enzymes are used in many industrial processes. One enzyme that stands out for its importance is β-glucosidase (β-D-glucohydrolase; E.C. 3.2.1.21). This enzyme is present in many living beings such as microbes, plants, and mammals catalyzing many reactions [38]. Related to food processing and industry, this enzyme holds great importance in cellulose degradation for it is responsible for glucose liberation as end-product from cellulose [39]. However, for achieving its hydrolytic activity, other enzymes must be present and its degradation products, such as cello-oligosaccharides and cellobiose [40]. These enzymes are endo β–1,4-glucanase (E.C. 3.2.1.4) and exo β–1,4-glucanase or cellobiohydrolase (E.C. 3.2.1.91). Cellulose, combined with hemicellulose and lignin, constitutes lignocellulose, which is the most abundant polymer in our planet [41]. The β-glucosidase enzyme is susceptible to inhibition if products from the previously mentioned enzymes accumulates, thus, stopping complete cellulose hydrolysis [42]. So, it can be said that β-glucosidase is the most important enzyme in cellulose degradation from an operational point of view [43]. Another interesting activity that β-glucosidase can carry out is degradation of glycosylated flavonoids that have influence on wine affecting flavor and aroma [14, 44]. Some of the compounds liberated by β-glucosidase are bioactive [45]. On the other hand, β-glucosidase can manifest transglycosylase activity, which have a ''reverse effect','' synthesizing glycosides and oligosaccharides [46, 47]. Both catalytic activities are shown in Figure 3.1. These aspects are of great interest in the food industry by its organoleptic effect and for high added-value compounds addition. These enzymes do not have a defined classification among its group, it has been separated as its substrate specificity and its nucleotide sequence [48]. While β-glucosidase is associated to the final step of cellulose degradation, its capability of action on many compounds, makes β-glucosidase a highly versatile enzyme regarding food industry.

### 3.3.1  PRODUCTION OF β-GLUCOSIDASE UNDER SOLID-STATE CONDITIONS

Many studies associated with β-glucosidase production turn to explore the effect of solid-state conditions [49, 50]. In table 3.1 are shown several examples of microbes that are able to produce β-glucosidase in a solid-state fermentation process. According to several studies, β-glucosidase could take extended periods of time to reach its maximum activity in the bioprocess [21, 51]. Even though its production is successful, for better results, shorter periods of production are more suitable if solid-state

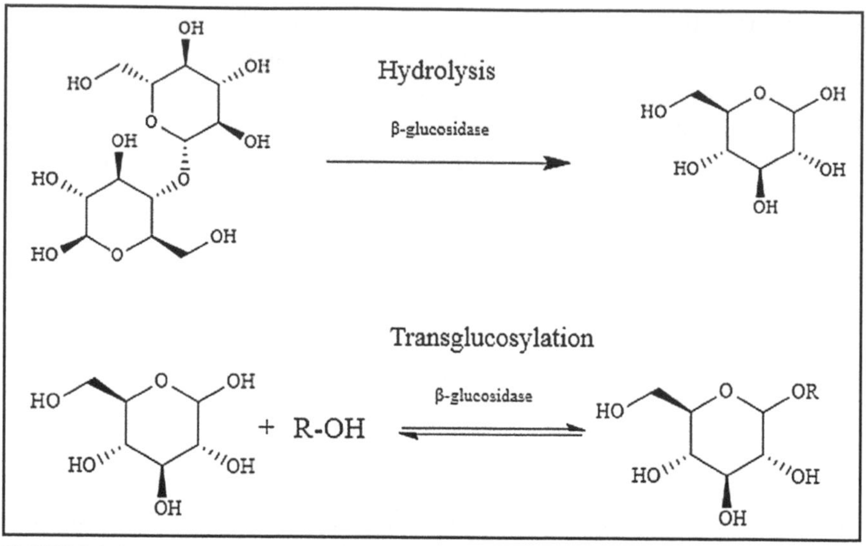

**FIGURE 3.1**   General scheme of the enzymatic activities of β-glucosidase.

systems are to be developed. Some research groups are reporting shorter production periods of β-glucosidase [52, 53]. Revising their results, there is a possible explanation for improving enzyme production periods, which could be the presence of non-cellulosic compounds together with cellulose, or even without cellulose presence [54]. Considering the substrates, the composition is the most important aspect for the expected results. Several substrates with high cellulose content have been used in solid-state bioprocesses such as agave leaf fibers [21], wheat bran [40], cranberry pomace [51], corncob [55], among others. These studies had as a result production periods ranging from 72 hours to 20 days. Although successful, other works report shorter enzyme production periods such as pomegranate polyphenols [7] and tarbush leaves [52] fermentation in solid-state conditions, where the highest β-glucosidase activity was registered at 24 and 36 hours, with 525 and 4862 U/L respectively. Also, using substrates such as corncob, fungal strains such as *Aspergillus niger* can produce up to 6436 U/L in 24 hours [55] while *Trichoderma harzianum* produces 2500 U/L at 96 hours [53]. By using corncob, a phenomenon could take place, considering the work of Lu et al. [56], which estates that in the presence of xylose as sole carbon source, *Aspergillus niger* can produce, aside from xylosidases and xylanases, a wide array of hydrolases,

**TABLE 3.1** Examples of Microbial Production of β-Glucosidase Under Solid-State Fermentation

| Strain | Substrate | Enzymes | Production time | Reference |
|---|---|---|---|---|
| **Fungi** | | | | |
| Aspergillus niger | Several substrates | Endoglucanase, exoglucanase, and β-glucosidase | 72 h | [40] |
| Lentinus edodes | Cranberry pomace | β-glucosidase | 20 days | [51] |
| Trichoderma harzianum | Corncob | Endoglucanase, exoglucanase, and β-glucosidase | 96 h | [55] |
| Aspergillus niger | Flourensia cernua foliage | β-glucosidase | 36 h | [88, 89] |
| Trichoderma asperellum | Agave fibers | Endoglucanase, exoglucanase, and β-glucosidase | 300 hours | [21] |
| Lichtheimia ramosa | Several substrates | β-glucosidase | 96 h | [50] |
| Penicillium citrinum | Rice bran | β-glucosidase | 96 h | [90] |
| Aspergillus and Trichoderma | Flax fiber | β-glucosidase among other hydrolytic enzymes | 4–11 days | [91] |
| Aspergillus niger | Pomegranate extract | β-glucosidase among other hydrolytic enzymes | 24 hours | [54] |
| **Bacteria** | | | | |
| Bacillus sp | Soybean | β-glucosidase and α-amylase | 24 hours | [92] |
| Lactobacillus paracasei | Soy | β-glucosidase and α-galactosidase | 24 hours | [93] |
| Bifidobacterium longum | Soy | β-glucosidase and α-galactosidase | 24 hours | [93] |
| **Yeast** | | | | |
| Saccharomyces cerevisiae, Hanseniaspora valbyensis and Hanseniaspora uvarum | Apple pomace | β-glucosidase among other enzymes | 4 weeks | [94] |

including β-glucosidase. Since corncob is known for its xylan content, that could promote β-glucosidase production in short periods of time. Another example of β-glucosidase production is the already mentioned work of Ascacio-Valdés et al. [54], where the substrate was composed only by pomegranate ellagitannin, inoculated with *Aspergillus niger*, and among other hydrolases, β-glucosidase was produced. In a work reported by Huerta-González [52], tarbush was fermented in a solid bioprocess with *A. niger* where the highest β-glucosidase activity was obtained at 36 hours of culture. In these two examples, there could be a close relation between cellulose, hemicellulose, and glycosides that could induce β-glucosidase production. Our research group is currently evaluating the interaction of tarbush composition before and after fungal fermentation, while taking into consideration the presence of phenolic glycosides in β-glucosidase production.

### 3.3.2  β-GLUCOSIDASE ACTIVITIES

#### 3.3.2.1  HYDROLASE ACTIVITY

The hydrolytic activity of β-glucosidase is very highly valued action in several industrial interests. In one hand, the biofuel industry, many of which utilizes residues from agricultural and food industry. On the other hand, the cleavage of glycosidic compounds that have organoleptic effect on food products has high importance. β-glucosidase hydrolyses β–1,4 glycosidic bonds in cellobiose, which is a dimer of two glucose molecules that possess the aforementioned bond. Another way that β-glucosidase activity can manifest is the hydrolysis of glycosidic compounds such as phenolic glycosides. These compounds are located in vegetal tissue such as ellagitannins, gallotannins, flavonol glycosides, among others [57–59]. The most common way to evaluate β-glucosidase activity is the use of a spectrophotometric method, which involves *p*-nitrophenyl-β-D-glucopyranoside, from which, if the activity is present, releases *p*-nitro-phenol and glucose, being the first that is measured at 500 nm of absorbance [51]. Another very common activity method is the use of cellobiose as substrate which is measured as reducing sugar released by the enzyme. If a chromatographer is available, the enzymatic reaction can measure glucose from a cellobiose reaction assay using a column that separates sugars such as an Aminex HPX–87P, along with a refractive index detector [60]. As the cleavage of cellobiose is the most known hydrolytic activity, there are other reactions that β-glucosidase is capable of catalyzing. As previously said, this enzyme can cleave sugar from non-sugar molecules such as glycosylated phenolics. There are several compounds related to industrial food processes, such as naringin where this compound gives a bitter flavor in citrus. An enzyme called naringinase (3.1.2.40), displays both α-rhamnosidase and β-glucosidase that first releases rhamnose followed by glucose and the respective aglycone [14]. Another example of hydrolytic activity in this type of compound is the hydrolysis of dhurrin, which is a cyanogenic glucoside, which is found in sorghum malt. In this case, β-glucosidase cleaves the glycosidic bond to release glucose and hydroxymandelonitrile [61]. Compounds that have bioactivities such as antioxidants are abundant. The case of tannins, specifically gallotannins and ellagitannins, are gallic and ellagic acid moieties bonded to a polyol core, which is commonly a sugar such as glucose [62, 63]. There have been reports that states that β-glucosidase can hydrolyse the molecule [51] and that those types of molecules can induce fungal β-glucosidase production [54]. There is also evidence that β-glucosidase can hydrolyze glycosylated precursors of terpenes and glucosylceramide releasing volatile terpenes and ceramides [64].

### 3.3.2.2   TRANSGLYCOSYLATION ACTIVITY

β-glucosidase is also capable to bind molecules together. This activity transfer sugar molecules with other sugars or other types of compounds. There are many transglucosylases in nature produced by mammals, microorganisms or plants [38, 65]. The main interest in this case is the transglycosylation activity of β-glucosidase and also transglycosylation because it can affect molecules that contain sugars different from glucose. In this enzymatic reaction, two main reagents are needed for it to occur; a donor molecule and an acceptor molecule (Figure 3.2). The common donor molecule is a sugar where a hydroxyl group is available to react with an acceptor molecule which can be another sugar or any different molecules that can bond with a sugar [66]. The most common product of transglycosylation of a β-glucosidase is oligosaccharides. In this case, for the formation of transglycosylation products, there are reports of a high-tolerance β-glucosidase which also exhibited said activity by producing cellobiose, cellotriose, sophorose, gentibiose, and laminaribiose [67]. These synthesis processes are favored because they display higher stereo and regio-selectivity by using β-glucosidase from *Aspergillus niger* and cellobiose as substrate, oligomers of β–1,4 and β–1,6 [68]. There is another form of glycosylation to generate other types of compounds. One of the most abundant kinds of glycosylated molecules are phenolic compounds. Naturally these molecules can be found in plants in the form gallotannins, ellagitannins, glycosylated cinnamic acids, flavonoids, flavanones, antocyanins, terpenes or chalcones, to name a few [5, 62]. These molecules can be obtained via enzyme pathway by transglycosylases, which can also have microbiological origin. Many microorganisms have been reported to produce β-glucosidases, some of which are fungal strains, which in theory, the enzyme of interest could be produced by solid-state cultures given the fact that fungi can fully adapt to solid-state growth conditions. Examples of glucosylation have targeted molecules such as dihydromyricetin, hespererin, quercetin, kaempferol by several microorganisms such as *Trichoderma*, *Aspergillus*, *Penicillium*, and *Rhizopus* [1, 69].

The preferred process of transglycosylation or reverse hydrolysis is the production of oligosaccharides [70]. In this case, for the reaction requires

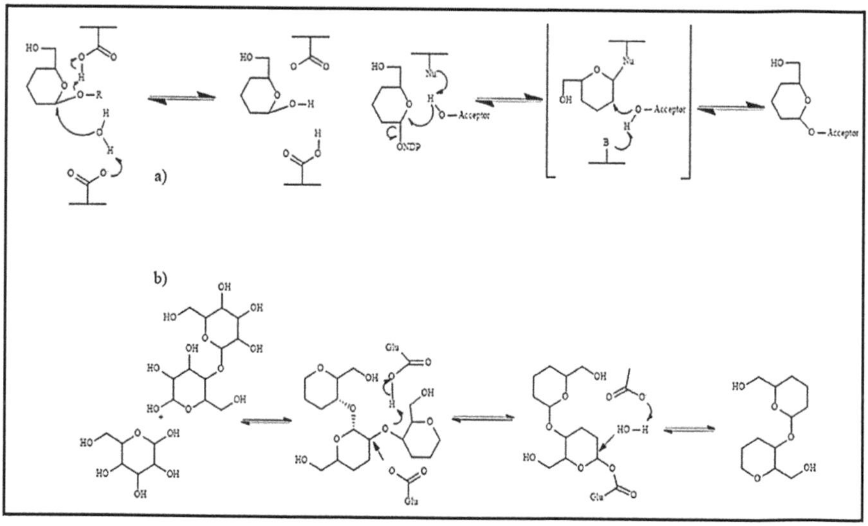

**FIGURE 3.2** Reaction mechanism of enzymatic transglycosylation (a) and reverse hydrolysis (b).

the presence of glucose so they can be polymerized which is considered more economically viable compared with alkyl or aryl-glycosylation. The synthesis of glucosides different from oligosaccharides requires nucleotide-activated sugar donors which are expensive [71]. Aside from the drawbacks of this type of reaction, enzymatic glycoside synthesis is now a "boom" regarding green chemistry with a very ample field of improving the viability of glycoside synthesis [15].

### 3.3.3   USE OF β-GLUCOSIDASE PRODUCTS IN AND RELATED TO FOOD INDUSTRY

As many enzymes have many uses in food processing, β-glucosidase possess an apparent action with a more refined objective. Some of the food products that use β-glucosidase are fruits, vegetables, teas, wine, and flavor compounds. For example, enzyme plays an important role in flavor improvement in wine and beer. The organoleptic features of fermented products are what define the acceptance of the consumer. The versatility of β-glucosidase that includes the hydrolytic and transglycosylation activities, gives the enzyme high importance in certain types of food and its processes.

#### 3.3.3.1   HYDROLYTIC ACTIVITY

Several compounds of interest in the food industry can be hydrolyzed by β-glucosidase (Figure 3.3). Many of the molecules responsible to add favorable organoleptic features to the wine, are glycosylated compounds. In winemaking, pectinase, protease, tannase, and β-glucosidase are used to improve product quality. For aroma compound release, terpenes are attributed to contributing in this matter as glucosylated precursors. β-glucosidase can hydrolyze glucosylated terpenes to release the volatile compound to add aroma to the final product [72]. Even though *Saccharomyces cerevisiae* is known to produce β-glucosidase, it is very common that the enzyme production is absent during fermentation, possibly because the growth conditions do not favor β-glucosidase production by the yeast [73]. For improvement of wine quality, glycosidases have been studied in aroma release in cherry wine [74]. Among those glycosidases is β-glucosidase.

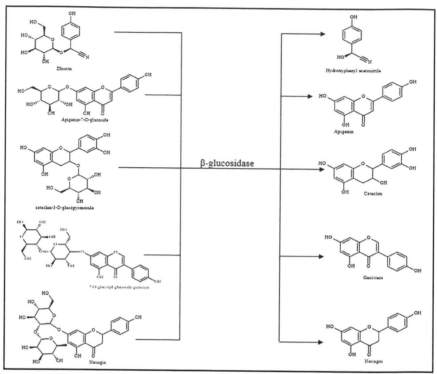

**FIGURE 3.3**　Glucosides present in food products that can be degraded by β-glucosidase with their respective aglycones.

As previously mentioned, β-glucosidase can free terpenes from its glyco-sylated precursors.

Enzymatic processes are developed to improve flavor in food. One example is the debittering process of citrus fruits. To achieve this purpose, an enzyme called naringinase can effectively degrade bittering compounds such as naringin. This enzyme has two active sites, which possesses both α-rhamnosidase and β-glucosidase activities. In this case, by breaking down the molecule it releases rhamnose, leaving prunin as a resulting molecule where β-glucosidase will release a glucose monomer and an aglycone, thus, contributing decreasing bitterness in citrus fruits [14].

Soybean is widely known for its protein content and nutritional value, but also for containing isoflavones with bioactivity. It possesses β-glycosides, which can be found among isoflavones. Some of these compounds are the aglycone form of genistein, daidzin, and glycitein, which have sugars bonded with them. β-glucosidase has the capacity to break down the molecules and release the aglycones, which have importance in food industry [75, 76]. In this process, the nutritional value of soybean can be improved by providing isoflavones of easier availability with higher bioactivity than its glycosylated precursors. There is a specific case of β-glucosidase production under solid-state fermentation using soy flour [76]. Among the results, the fungal strain of *Monascus purpureus* produced lower levels of β-glucosidase than the other used strain of *Aspergillus oryzae*. The advantage offered by *M. purpureus* is that aglycones of high-added value were present in higher amounts in its solid-state cultures. It is worth mentioning that both fungi are used to produce fermented food. These results provide another advantage, which is the evidence that a substrate with free aglycone (isoflavones) presence increase the nutritional and functional value of fermented soy-based products that can be partially attributed to β-glucosidase activity.

There is a beer making process in Africa that is made from sorghum malt which its organoleptic characteristics are affected by the presence of dhurrin. This glucoside is present in low quantities in sorghum seed [77, 78]. For the production of the beer, malting of the sorghum is carried out to activate enzymes that degrade proteins and starch. Presence of β-glucosidase may prove difficult to unify the activity of the rest of the enzymes due to different optimal conditions, so special care is required to efficiently degrade components for beer production and flavor compounds adequate accumulation. β-glucosidase plays an important role in releasing glucose from the cyanogenic fraction of the molecule, which is 4-hydroxyphenyl acetonitrile. The presence of microbial origin β-glucosidase has promoted the reduction of dhurrin quantity in beer and increases the release of flavor compounds improving organoleptic quality. The hydrolysis can promote the formation of hydrogen cyanide which is highly toxic. There is an advantage that malting requires heating and hydrogen cyanide, being highly volatile, can be removed prior to fermentation. However, the poor quality control of raw material promotes the addition of germinated sorghum in the process which contains high amounts of dhurrin. This compound, in higher quantities, can hinder beer quality and could lead to the formation of ethyl carbamate in alcoholic fermentation by the reaction with ethanol and hydrogen cyanate [61].

Bacteria have also been used in solid-state bioprocesses [79] and for β-glucosidase production [80]. Anaerobic conditions have been established by the growth of *Bacillus, Bifidobacterium,* and *Lactobacillus* strains for whole soybean fermentation [81] where β-glucosidase could be approached under these conditions for a more active and stable enzyme. Important studies that include lactic acid bacteria and β-glucosidase are the breakdown of molecules such as oleuropein. These molecules are found in olives, where a conservation process involves lactic fermentation, although the bioprocess does not depend on the hydrolysis of the molecule. Bacteria from this bioprocess have been reported to break down the molecule which promotes bitterness reduction in olives [67]. Lactic acid bacteria have also been reported to increase isoflavones such as genistein and daidzin by fermenting soy milk [82].

## 3.3.3.2   *TRANSGLUCOSYLASE ACTIVITY*

Using β-glucosidases for chemical synthesis is a very promising technology and holds great research interest for using a biocatalyst in green chemistry. Several types of compounds have been synthesized by these means such as vitamin, alkyl, stilbenoids, flavonoid glycosides or oligosaccharides. Reversing the hydrolysis to form more complex molecules by β-glucosidase is an enzymatic reaction that can have applications in pharmaceutical and food industries [15, 83]. As β-glucosidases are mostly known as saccharide cleavers, certain conditions must be met to promote glycosylation reactions. For starters, the two activities can be present, thus, if hydrolytic activity is faster, transglycosilation may not be detected. For achieving a higher rate of transglycosylation, the use of high substrate concentration and lower water activity can be implemented [70]. There are other pathways to achieve transglycosylation products, but require expensive reagents. In this case, a glycosyltransferase could be used, but the presence of an activator such as UDP-Glucose (Uracil diphosphate glucose) is needed as an electron donor to form bonds with another sugar [17]. As previously mentioned, this molecule is very expensive and it represents bottleneck in transferase products. Having an enzyme such as β-glucosidase that is able to reproduce the same type of reaction products is an advantage because it is of lower cost and higher accessibility because it can be obtained from microbes. There are many glycosylation reactions, from which the most used in food industry are the following:

### 3.3.3.2.1   Alkyl-β-D-glucosides

The synthesis product of this reaction is the union of two molecules, being one a glucose molecule and an alkyl group, which is an alkane substituent (Figure 3.4). These molecules are used as biodegradable non-ionic surfactants with a wide range of applications. The molecules required for the enzymatic reaction to take place is a glucose and a primary alcohol, which can be of different length and also can be saturated or unsaturated [15, 71, 84]. The usual production of these types of molecules obeys the following mechanism. In this reaction, an electron donor and acceptor molecules must be present. These reagents are glycosyl donors and glycosyl acceptors with β-glucosidase as the catalyzer where several alkyl-β-glucosides can be produced.

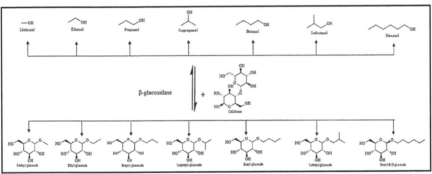

**FIGURE 3.4**   Examples of alkyl-glucosides that can be produced by transglycosylation by β-glucosidase.

### 3.3.3.2.2   Aryl-β-D-Glucosides

When aromatic compounds react to form bonds with glycones, Aryl-β-D-glycosides are formed. There are enzymatic reactions which include β-glucosidases to synthesize these compounds (Figure 3.5). Addressing the mechanism, it follows the same type of reagent molecules as the alkyl-glucosides synthesis. Many of these molecules are found in nature as part of the composition of plant tissue but processes have been developed to synthesize these compounds by biotechnological way for its importance in several food processes [69]. Such compounds can be associated with isoflavones. The aglycone derivates can have low water solubility, thus by the action of a binding enzyme, these molecules can be glucosylated becoming water-soluble and bioactivities are improved. Glycosylation of bioactive phenolic compounds can have the limitation of electron donors such as nucleotide-activated donors. In the case of aryl-β-D-glucosides of food industry interest, compounds of high added-value are synthesized such as isoflavones or other phenolic compounds of interest. Although β-glucosidases are reference enzymes to produce aryl-β-D-glucosides, there are reports of β-glycosidases, which involve β-glucosidases, to synthesize aryl-β-D-glucosides by yeasts [71].

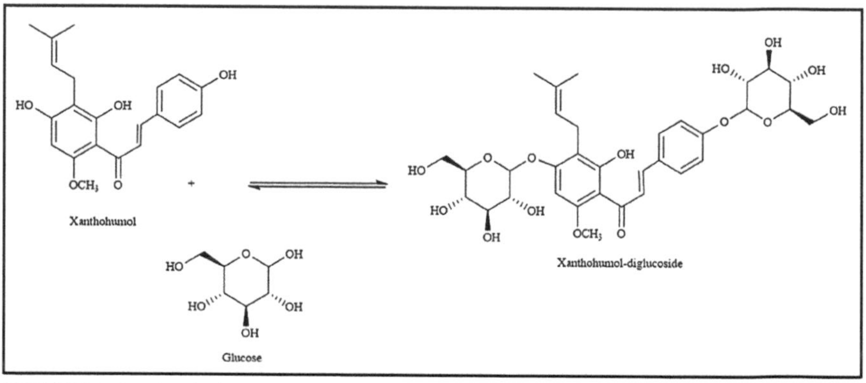

**FIGURE 3.5**    Aryl-glucoside produced by transglycosylation activity of a β-glucosidase.

## 3.4   OLIGOSACCHARIDES PRODUCTION

The synthesis of oligosaccharides is the process of a more adequate application in food industry, mainly in probiotic inclusion in food products (Figure 3.6). As previously discussed, for glycosidic bonds to be created, β-glucosidase is able to bring synthetic activity in its reactions. The synthetic reactions can take two ways which are reverse hydrolysis or transglycosylation. The requirements for reverse hydrolysis to take place, instead of a hydrolysis, is the lowering of water activity, high substrate concentration and product trapping. These actions could shift the reaction pattern into forming glycosidic bonds instead of a cleavage reaction. It is possible that the mechanism of hydrolysis of cellulose is slowed down or inhibited due to the accumulation of cellobiose which is attributed to transglycosylation activity of β-glucosidase, which in turn, the product of reaction binds to the enzyme, slowing or stopping the hydrolysis rate. There are reports of oligomer production by fungal β-glucosidases from several genera including *Aspergillus* and *Penicillium* [68]. The fungal strain of *Talaromyces termophilus* is able to produce cellobiose, cellotriose, cellotetraose, and cellopentaose, also from by means of a β-glucosidase. Other oligosaccharides have been synthesized by β-glucosidase such as gentibiose, laminariabiose, sophorose, and gentibiose [85]. Cello-oligosaccharides or β-glucans have been reported to favor certain functions and features in the human body, such as lowering cholesterol levels, improves blood pressure and lipid levels, contributes to the betterment of glycemic index and insulin function and helps in maintaining an adequate diet and helps lowering weight [86, 87].

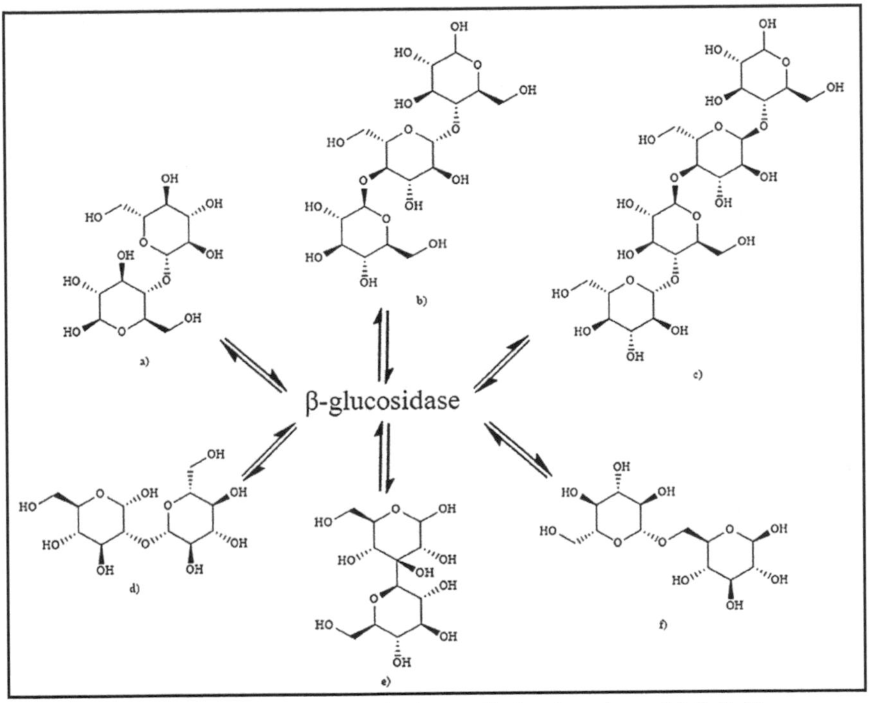

**FIGURE 3.6** Enzymatic synthesis of β-glycosydic-bond products. (a) Cellobiose (β–1,4); (b) Cellotriose (β–1,4); (c) Cellotetraose (β–1,4); (d) Sophorose (β–1,2); (e) Laminaribiose (β–1,3); and (f) Gentiobiose (β–1,6).

## 3.5  CONCLUDING REMARKS

The versatility of manifesting extra activities are important aspects of enzymes to be taken into account while bioprocessing materials. Solid-state fermentation, being a process that is generally carried out using agro-industrial residues, is a common source of β-glucosidase, mostly of fungal origin. Regardless of the means of production, β-glucosidase in its hydrolytic and transglycosylic activity represents a very important asset to the food industry. The release of glucose from cellulose is the most known activity of β-glucosidase. The cleavage of glycosylated compounds and thus modifying organoleptic characteristics of food is a very important reaction. The synthesis of molecules using β-glucosidase also has high impact in food industry, where alkyl-glucosides can be used as surfactants with applications in food processes. Another strong product of β-glucosidase as synthesis catalyzer is the production of oligosaccharides with plenty of use in food industry as part of dietary fibers or as growth promoters of probiotics.

## KEYWORDS

- **food industry**
- **transglycosylation**

- **β-glucosidase**

## REFERENCES

1. Lin, S., Yang, B., Chen, F., Jiang, G., Li, Q., Duan, X., et al., (2012). Enhanced DPPH radical scavenging activity and DNA protection effect of litchi pericarp extract by *Aspergillus awamori* bioconversion. *Chem. Cent. J.*, *6*(1), 108.
2. Couto, S. R., & Angeles, S. M., (2006). Application of solid-state fermentation to food industry-A review. *J. Food Eng.*, *76*(3), 291–302.
3. Fernandes, P., & Carvalho, F., (2017). Microbial enzymes for the food industry. *Biotechnology of Microbial Enzymes Biotechnology of Microbial Enzymes*, Academic Press. Chapter 19, 531–544.

4. Ghorai, S., Banik, S. P., Verma, D., Chowdhury, S., Mukherjee, S., & Khowala, S., (2009). Fungal biotechnology in food and feed processing. *Food Res. Int.*, *42*(5–6), 577–587.

5. Martins, S., Mussatto, S. I., Martínez-Avila, G., Montañez-Saenz, J., Aguilar, C. N., & Teixeira, J. A., (2011). Bioactive phenolic compounds: Production and extraction by solid-state fermentation. A review. *Biotechnol. Adv.*, *29*(3), 365–373.

6. Dulf, F. V., Vodnar, D. C., & Socaciu, C., (2016). Effects of solid-state fermentation with two filamentous fungi on the total phenolic contents, flavonoids, antioxidant activities and lipid fractions of plum fruit (*Prunus domestica* L.) by-products. *Food Chem.*, *209*, 27–36.

7. Ascacio-Valdés, J. A., Aguilera-Carbó, A. F., Buenrostro, J. J., Prado-Barragán, A., Rodríguez-Herrera, R., & Aguilar, C. N., (2016). The complete biodegradation pathway of ellagitannins by *Aspergillus niger* in solid-state fermentation. *J. Basic. Microbiol.*, *56*, 329–336.

8. Soccol, C. R., Scopel, E., Alberto, L., Letti, J., Karp, S. G., & Woiciechowski, A. L., (2017). Recent developments and innovations in solid-state fermentation. *Biotechnol. Res. Innov.*, 1–20. doi:10.1016/j.biori.2017.01.002.

9. Singhania, R. R., Patel, A. K., Soccol, C. R., & Pandey, A., (2009). Recent advances in solid-state fermentation. *Biochem. Eng. J.*, *44*(1), 13–18.

10. Pandey, A., (2003). Solid-state fermentation. *Biochem. Eng. J.*, *13*(2), 81–84.

11. Gao, L., Gao, F., Jiang, X., Zhang, C., Zhang, D., Wang, L., Wu, G., & Chen, S., (2014). Biochemical characterization of a new β-glucosidase (Cel3E) from *Penicillium piceum* and its application in boosting lignocelluloses bioconversion and forming disaccharide inducers: New insights into the role of β-glucosidase. *Process Biochem.*, *49*(5), 768–774.

12. Sorensen, A., Andersen, J. J., Ahring, B. K., Teller, P. J., & Lubeck, M., (2014). Screening of carbon sources for β-glucosidase production by *Aspergillus saccharolyticus*. *Int. Biodeterior. Biodegrad.*, *93*, 78–83.

13. Boudabbous, M., Ben Hmad, I., Saibi, W., Mssawra, M., Belghith, H., & Gargouri, A., (2017). Trans-glycosylation capacity of a highly glycosylated multi-specific β-glucosidase from *Fusarium solani*. *Bioprocess Biosyst. Eng.*, *40*(4), 559–571.

14. Cui, P., Li, T. D. S., Ping, L. Z., & Sun, W. Y., (2016). Highly selective and efficient biotransformation of linarin to produce tilianin by naringinase. *Biotechnol. Lett.*, *38*(8), 1367–1373.

15. Kumar, P., Ryan, B., & Henehan, G. T. M., (2017). β-Glucosidase from *Streptomyces griseus*: Nanoparticle immobilization and application to alkyl glucoside synthesis. *Protein Expr. Purif.*, *132*, 164–170.

16. Pei, X., Zhao, J., Cai, P., Sun, W., Ren, J., Wu, Q., et al., (2016). Heterologous expression of a GH3 β-glucosidase from *Neurospora crassa* in *Pichia pastoris* with high purity and its application in the hydrolysis of soybean isoflavone glycosides. *Protein Expr. Purif.*, *119*, 75–84.

17. Pandey, R. P., Parajuli, P., Koffas, M. A. G., & Sohng, J. K., (2016). Microbial production of natural and non-natural flavonoids: Pathway engineering, directed evolution and systems/synthetic biology. *Biotechnol. Adv.*, *34*(5), 634–662.

18. Ambigaipalan, P., Costa De Camargo, A., Shahidi, F., Priyatharini, A. M. N., & De Camargo, A. C., (2016). Identification of phenolic antioxidants and bioactives of pomegranate seeds following juice extraction using HPLC-DAD-ESI. *Food Chem.*, *221*, 1883–1894.

19. Singhania, R. R., Sukumaran, R. K., Patel, A. K., Larroche, C., & Pandey, A., (2010). Advancement and comparative profiles in the production technologies using solid-state and submerged fermentation for microbial cellulases. *Enzyme Microb. Technol.*, *46*(7), 541–549.

20. Medina, M. A., Belmáres, R. E., Aguilera-Carbo, A., Rodríguez-Herrera, R., Aguilar, C. N., Fungal culture systems for production of antioxidant phenolics using pecan nut shells as sole carbon source. *Am. J. Agric. Biol. Sci.*, *5*(3), 397–402.

21. Nava-Cruz, N. Y., Contreras-Esquivel, J. C., Aguilar-González, M. A., Nuncio, A., Rodríguez-Herrera, R., & Aguilar, C. N., (2016). *Agave atrovirens* fibers as substrate and support for solid-state fermentation for cellulase production by *Trichoderma asperellum*. *3 Biotech.*, *6*(1), 115.

22. Pereira, F., Almeida, D., Maria, D., Freire, G., & Lins, U., (2017). Surface imaging of the filamentous fungus *Penicillum simplicissimum* growing in a solid-state fermentation system. *Micron.*, *99*, 19–25.

23. Yoon, L. W., Ang, T. N., Ngoh, G. C., & Chua, A. S. M., (2014). Fungal solid-state fermentation and various methods of enhancement in cellulase production. *Biomass Bioenergy.*, *67*, 319–338.

24. Da Silva Delabona, P., Lima, D. J., Robl, D., Rabelo, S. C., Farinas, C. S., & da Cruz Pradella, J. G., (2016). Enhanced cellulase production by *Trichoderma harzianum* by cultivation on glycerol followed by induction on cellulosic substrates. *J. Ind. Microbiol. Biotechnol.*, *43*(5), 617–26.

25. Ahamed, A., & Vermette, P., (2008). Enhanced enzyme production from mixed cultures of *Trichoderma reesei* RUT-C30 and *Aspergillus niger* LMA grown as fed batch in a stirred tank bioreactor. *Biochem. Eng. J.*, *42*(1), 41–46.

26. Gupta, V. K., Kubicek, C. P., Berrin, J. G., Wilson, D. W., Couturier, M., Berlin, A., et al., (2016). Fungal enzymes for bio-products from sustainable and waste biomass. *Trends Biochem. Sci.*, *41*(7), 633–645.

27. Kumar, R., Singh, S., & Singh, O. V., (2008). Bioconversion of lignocellulosic biomass: Biochemical and molecular perspectives. *J. Ind. Microbiol. Biotechnol.*, *35*(5), 377–391.

28. Sánchez, C., (2009). Lignocellulosic residues: Biodegradation and bioconversion by fungi. *Biotechnol. Adv.*, *27*(2), 85–194.

29. Liang, J., Peng, X., Yin, D., Li, B., Wang, D., & Lin, Y., (2015). Screening of a microbial consortium for highly simultaneous degradation of lignocellulose and chlorophenols, *Bioresour. Technol.*, *190*, 381–387.

30. Arévalo-Gallegos, A., Ahmad, Z., Asgher, M., Parra-Saldivar, R., & Iqbal, H. M., (2017). Lignocellulose : A sustainable material to produce value-added products with a zero waste approach- A review. *Int. J. Biol. Macromol.*, *99*, 308–318.

31. Graminha, E. B. N., Gonzalves, A. Z. L., Pirota, R. D. P. B., Balsalobre, M. A. A., Da Silva, R., & Gomes, E., (2008). Enzyme production by solid-state fermentation: Application to animal nutrition. *Anim. Feed Sci. Technol.*, *144*(1), 1–22.

32. Nitayapat, N., Prakarnsombut, N., Lee, S. J., & Boonsupthip, W., (2015). Bioconversion of tangerine residues by solid-state fermentation with *Lentinus polychrous* and drying the final products. *LWT – Food Sci. Technol., 63*(1), 773–779.

33. Aires, A., Carvalho, R., & Saavedra, M. J., (2016). Valorization of solid wastes from chestnut industry processing: Extraction and optimization of polyphenols, tannins and ellagitannins and its potential for adhesives, cosmetic and pharmaceutical industry. *Waste Manag., 48*, 457–464.

34. Melikoglu, M., Lin, C. S. K., & Webb, C., (2013). Stepwise optimisation of enzyme production in solid-state fermentation of waste bread pieces. *Food Bioprod. Process, 91*(4), 638–646.

35. Pensupa, N., Jin, M., Kokolski, M., Archer, D. B., & Du, C., (2013). A solid-state fungal fermentation-based strategy for the hydrolysis of wheat straw. *Bioresour. Technol., 149*, 261–267.

36. Aggelopoulos, T., Katsieris, K., Bekatorou, A., Pandey, A., Banat, I. M., & Koutinas, A. A., (2014). Solid-state fermentation of food waste mixtures for single cell protein, aroma volatiles and fat production. *Food Chem., 145*, 710–716.

37. Ang, S. K., Shaza, E. M., Adibah, Y. A., Suraini, A. A., & Madihah, M. S., (2013). Production of cellulases and xylanase by *Aspergillus fumigatus* SK1 using untreated oil palm trunk through solid-state fermentation. *Process Biochem., 48*(9), 1293–1302.

38. Ketudat, C. J. R., Mahong, B., Baiya, S., & Jeon, J. S., (2015). β-Glucosidases: Multitasking, moonlighting or simply misunderstood?. *Plant Sci., 241*, 246–259.

39. Pryor, S. W., & Nahar, N., (2015). β-glucosidase supplementation during biomass hydrolysis: How low can we go? *Biomass and Bioenergy, 80*, 298–302.

40. Sukumaran, R. K., Singhania, R. R., Mathew, G. M., & Pandey, A., (2009). Cellulase production using biomass feed stock and its application in lignocellulose saccharification for bio-ethanol production. *Renew. Energy, 34*(2), 421–424.

41. Goh, C. S., Tan, K. T., Lee, K. T., & Bhatia, S., (2010). Bio-ethanol from lignocellulose: Status, perspectives and challenges in Malaysia. *Bioresour. Technol., 101*(13), 4834–4841.

42. Andric, P., Meyer, A. S., Jensen, P. A., & Dam-Johansen, K., (2010). Reactor design for minimizing product inhibition during enzymatic lignocellulose hydrolysis: I. Significance and mechanism of cellobiose and glucose inhibition on cellulolytic enzymes. *Biotechnol. Adv., 28*(3), 308–324.

43. Abdella, A., Mazeed, T. E. S., El-Baz, A. F., & Yang, S. T., (2016). Production of β-glucosidase from wheat bran and glycerol by *Aspergillus niger* in stirred tank and rotating fibrous bed bioreactors. *Process Biochem., 51*(10), 1331–1337.

44. Singh, G., Verma, A. K., & Kumar, V., (2016). Catalytic properties, functional attributes and industrial applications of β-glucosidases. *3 Biotech., 6*(1), 1–14.

45. Rekha, C. R., & Vijayalakshmi, G., (2011). Isoflavone phytoestrogens in soymilk fermented with β-glucosidase producing probiotic lactic acid bacteria. *Int. J. Food Sci. Nutr., 62*(2), 111–120.

46. Shimba, N., Shinagawa, M., Hoshino, W., Yamaguchi, H., Yamada, N., & Ichiro Suzuki, E., (2009). Monitoring the hydrolysis and transglycosylation activity of β-glucosidase from *Aspergillus niger* by nuclear magnetic resonance spectroscopy and mass spectrometry. *Anal. Biochem., 393*(1), 23–28.

47. Lundemo, P., Karlsson, E. N., & Adlercreutz, P., (2017). Eliminating hydrolytic activity without affecting the transglycosylation of a GH1 β-glucosidase. *Appl. Microbiol. Biotechnol.*, *101*(3), 1121–1131.

48. Bhatia, Y., Mishra, S., & Bisaria, V. S., (2002). Microbial β-glucosidases: Cloning, properties, and applications. *Crit. Rev. Biotechnol.*, *22*(4), 375–407.

49. Junior, A. B., Borges, D. G., Tardioli, P. W., & Farinas, C. S., (2014). Characterization of β -glucosidase produced by *Aspergillus niger* under solid-state fermentation and partially purified using MANAE-Agarose. *Biotechnol. Res. Int.*, *2014*, doi. org/10.1155/2014/317092.

50. Lisboa-Garcia, N. F., Da Silva Santos, F. R., Gonçalves, F. A., Da Paz, M. F., Fonseca, G. G., & Leite, R. S. R., (2015). Production of β-glucosidase on solid-state fermentation by *Lichtheimia ramosa* in agroindustrial residues: Characterization and catalytic properties of the enzymatic extract. *Electron. J. Biotechnol.*, *18*(4), 314–319.

51. Vattem, D. A., & Shetty, K., (2003). Ellagic acid production and phenolic antioxidant activity in cranberry pomace (*Vaccinium macrocarpon*) mediated by *Lentinus edodes* using a solid-state system. *Process Biochem.*, *39*(3), 367–379.

52. Huerta, I., (2015). *Chemical Characterization and Fungal Fermentation of Semidesert Plants With Animal Feed Potential*. Mexico: Antonio Narro Agrarian Autonomous University, Dissertation. 96 p.

53. García-Galindo, A., (2016). Fungal Production of industrial enzymes in solid fermentation on corn cobs and their recovery with polyelectrolytes. Autonomous University of Coahuila, Dissertation. 85 p.

54. Ascacio-Valdés, J. A., Buenrostro, J. J., De la Cruz, R., Sepúlveda, L., Aguilera-Carbó, A. F., Prado, A., et al., (2014). Fungal biodegradation of pomegranate ellagitannins. *J. Basic Microbiol.*, *54*, 28–34.

55. Gómez-García R., (2016). Fungal production of industrial enzymes (cellulases, xylanases and invertase) and study of its recovery with flexible-chain polymers. Autonomous University of Coahuila., Dissertation, 106 p.

56. Lu, X., Sun, J., Nimtz, M., Wissing, J., Zeng, A., & Rinas, U., (2010). The intra- and extracellular proteome of *Aspergillus niger* growing on defined medium with xylose or maltose as carbon substrate. *Microb. Cell Fact.*, *9*(1), 1–13.

57. Ma, Q., Xie, H., Li, S., Zhang, R., Zhang, M., & Wei, X., (2014). Flavonoids from the pericarps of *litchi chinensis. J. Agric. Food Chem.*, *62*(5), 1073–1078.

58. Ni, H., Chen, F., Jiang, Z. D., Cai, M. Y., Yang, Y. F., Xiao, A. F., et al., (2015). Biotransformation of tea catechins using *Aspergillus niger* tannase prepared by solid-state fermentation on tea byproduct. *LWT – Food Sci. Technol.*, *60*(2), 1206–1213.

59. Castro-López, C., Ventura-Sobrevilla, J. M., González-Hernández, M. D., Rojas, R., Ascacio-Valdés, J. A., Aguilar, C. N., et al., (2017). Impact of extraction techniques on antioxidant capacities and phytochemical composition of polyphenol-rich extracts. *Food Chem.*, *237*, 1139–1148.

60. Han, X., Song, W., Liu, G., Li, Z., Yang, P., & Qu, Y., (2017). Bioresource technology improving cellulase productivity of *Penicillium oxalicum* RE-10 by repeated fed-batch fermentation strategy. *Bioresour. Technol.*, *227*, 155–163.

61. Tokpohozin, S. E., Fischer, S., Sacher, B., & Becker, T., (2016). β-D-glucosidase as "key enzyme" for sorghum cyanogenic glucoside (dhurrin) removal and beer biofla-vouring. *Food Chem. Toxicol.*, *97*, 217–223.

62. Aguilera-Carbó, A., Augur, C., Prado-Barragan, L. A., Favela-Torres, E., & Aguilar, C. N., (2008). Microbial production of ellagic acid and biodegradation of ellagitan-nins. *Appl. Microbiol. Biotechnol.*, *78*(2), 189–199.

63. Valera, L. S., Jorge, J. A., & Guimarães, L. H. S., (2015). Characterization of a multi-tolerant tannin acyl hydrolase II from *Aspergillus carbonarius* produced under solid-state fermentation. *Electron. J. Biotechnol.*, *18*(6), 1–7.

64. Civeira, F., Recalde, D., García-Otín, A., & Cenarro, A., (2008). Genetic factors of cardiovascular disease. *International Encyclopedia of Public Health*. Academic Press, 44–55.

65. Mangas-Sánchez, J., & Adlercreutz, P., (2015). Enzymatic preparation of oligosac-charides by transglycosylation: A comparative study of glucosidases. *J. Mol. Catal. B Enzym.*, *122*, 51–55.

66. Mallek-Fakhfakh, H., & Belghith, H., (2016). Physicochemical properties of thermo-tolerant extracellular β-glucosidase from *Talaromyces thermophilus* and enzymatic synthesis of cello-oligosaccharides. *Carbohydr. Res.*, *419*, 41–50.

67. Rani, V., Mohanram, S., Tiwari, R., Nain, L., & Arora, A., (2014). Beta-Glucosidase : Key enzyme in determining efficiency of cellulase and biomass hydrolysis. *J. Biopro-cess Biotech.*, *5*(1), 1–8.

68. Bohlin, C., Praestgaard, E., Baumann, M. J., Borch, K., Praestgaard, J., Monrad, R. N., & Westh, P., (2013). A comparative study of hydrolysis and transglycosylation activities of fungal β-glucosidases, *Appl. Microbiol. Biotechnol.*, *97*(1), 159–169.

69. Xiao, J., Muzashvili, T. S., & Georgiev, M. I., (2014). Advances in the biotechnologi-cal glycosylation of valuable flavonoids. *Biotechnol. Adv.*, *32*(6), 1145–1156.

70. Ahmed, A., Nasim, F., Batool, K., & Bibi, A., (2017). Microbial β -glucosidase: Sources, production and applications. *J. Appl. Environ Microbiol.*, *5*(1), 31–46.

71. Rather, M., & Mishra, S., (2013). β-Glycosidases: An alternative enzyme based meth-od for synthesis of alkyl-glycosides. *Sustain. Chem. Process, 1*(1), 1–7.

72. Maicas, S., & Mateo, J., (2016). Microbial glycosidases for wine production. *Bever-ages, 2*(3), 1–11.

73. Daenen, L., Saison, D., Sterckx, F., Delvaux, F. R., VerachteErt, H., & Derdelinckx, G., (2008). Screening and evaluation of the glucoside hydrolase activity in *Saccharo-myces* and *Brettanomyces* brewing yeasts. *J. Appl. Microbiol.*, *104*(2), 478–488.

74. Wilkowska, A., & Pogorzelski, E., (2017). Aroma enhancement of cherry juice and wine using exogenous glycosidases from mould, yeast and lactic acid bacteria. *Food Chem.*, *237*, 282–289.

75. Maitan-Alfenas, G. P., Lage, L. G. D. A., De Almeida, M. N., Visser, E. M., De Rezende, S. T., & Guimarães, V. M., (2014). Hydrolysis of soybean isoflavones by *Debaryomyces hansenii* UFV-1 immobilised cells and free b –glucosidase. *Food Chem.*, *146*, 429–436.

76. Handa, C. L., Couto, U. R., Vicensoti, A. H., Georgetti, S. R., & Ida, E. I., (2014). Optimisation of soy flour fermentation parameters to produce β-glucosidase for bio-conversion into aglycones. *Food Chem.*, *152*, 56–65.

77. Whitfield, M. B., Chinn, M. S., & Veal, M. W., (2012). Processing of materials derived from sweet sorghum for biobased products. *Ind. Crops Prod.*, *37*(1), 362–375.

78. Tokpohozin, S. E., Julian, F. T. W., Fischer, S., & Becker, T., (2017). Polyphasic characterization of lactic acid bacteria isolated from Beninese sorghum beer starter. *LWT – Food Sci. Technol.*, *80*, 51–58.

79. Rai, A. K., Sanjukta, S., Chourasia, R., Bhat, I., Bhardwaj, P. K., & Sahoo, D., (2017). Production of bioactive hydrolysate using protease, β-glucosidase and α-amylase of *Bacillus* spp. isolated from kinema. *Bioresour. Technol.*, *235*, 358–365.

80. Michlmayr, H., & Kneifel, W., (2014). β-glucosidase activities of lactic acid bacteria: Mechanisms, impact on fermented food and human health. *FEMS Microbiol. Lett.*, *352*(1), 1–10.

81. Zhang, S., Shi, Y., Zhang, S., Shang, W., Gao, X., & Wang, H., (2014). Whole soybean as probiotic lactic acid bacteria carrier food in solid-state fermentation. *Food Control.*, *41*(1), 1–6.

82. Donkor, O. N., & Shah, N. P., (2008). Production of β-glucosidase and hydrolysis of isoflavone phytoestrogens by *Lactobacillus acidophilus*, *Bifidobacterium lactis*, and *Lactobacillus casei* in soymilk. *J. Food Sci.*, *73*(1), 15–20.

83. Thuan, N. H., & Sohng, J. K., (2013). Recent biotechnological progress in enzymatic synthesis of glycosides. *J. Ind. Microbiol. Biotechnol.*, *40*(12), 1329–1356.

84. De Roode, B. M., Franssen, M. C. R., Van Der Padt, A., & Boom, R. M., (2003). Perspectives for the industrial enzymatic production of glycosides. *Biotechnol. Prog.*, *19*(5), 1391–1402.

85. Uchiyama, T., Miyazaki, K., & Yaoi, K., (2013). Characterization of a novel β-glucosidase from a compost microbial metagenome with strong transglycosylation activity. *J. Biol. Chem.*, *288*(25), 18325–18334.

86. Daou, C., & Zhang, H., (2012). Oat Beta-Glucan : Its role in health promotion and prevention of diseases. *Compr. Rev. Food Sci. Food Saf.*, *11*, 355–365.

87. Chu, Q., Li, X., Xu, Y., Wang, Z., Huang, J., Yu, S., et al., (2014). Functional cello-oligosaccharides production from the corncob residues of xylo-oligosaccharides manufacture. *Process Biochem.*, *49*(8), 1217–1222.

88. López-Trujillo, J., Medina-Morales, M. A., Sánchez-Flores, A., Arévalo, C., Ascacio-Valdés, J. A., Mellado, M., et al., (2017). Solid bioprocess of tarbush (*Flourensia cernua*) leaves for β-glucosidase production by *Aspergillus niger*: Initial approach to fiber-glycoside interaction for enzyme induction. *3 Biotech.*, *7*, 271.

89. Medina-Morales, M. A., López-Trujillo, J., Gómez-Narváez, L., Mellado, M., García-Martínez, E., Ascacio-Valdés, J. A., et al., (2017). Effect of growth conditions on β-glucosidase production using *Flourensia cernua* leaves in a solid-state fungal bioprocess. *3 Biotech.*, *7*, 355.

90. Ng, I., Li, C., Chan, S., Chir, J., Chen, P., Tong, C., et al., (2010). High-level production of a thermoacidophilic β-glucosidase from *Penicillium citrinum* YS40-5 by solid-state fermentation with rice bran. *Bioresour. Technol.*, *101*, 1310–1317.

91. Szabo, O., Csiszar, E., Toth, K., Szakacs, G., & Koczka, B., (2015). Ultrasound-assisted extraction and characterization of hydrolytic and oxidative enzymes produced by solid-state fermentation. *Ultrason. Sonochem.*, *22*, 249–256.

92. Rai, A., Sanjukta, S., Chourasia, R., Bhat, I., Bhardwaj, P., & Sahoo, D., (2017). Production of bioactive hydrolysate using protease, β-glucosidase and α-amylase of *Bacillus* spp. Isolated from kinema. *Bioresour. Technol.*, 235, 358–365.
93. Selected lactobacilli and bifidobacteria development in solid-state fermentation using soybean paste, (2017). *Rev. Argent. Microbiol.*, 49(1), 62–69.
94. Rodríguez, R., Pando, R., & Suárez, B., (2015). Production and characterization of aroma compound from apple pomace by solid-state fermentation with selected yeasts. *LWT-Food Sci. Technol.*, 64, 1342–1353.

# CHAPTER 4

# PIGMENTED-GRAIN CORN IN MEXICO: IMPORTANCE AND POTENTIAL RISKS

MARÍA G. HERNÁNDEZ-ÁNGEL,[1]
JANETH MARGARITA VENTURA SOBREVILLA,[2]
DANIEL BOONE-VILLA,[3] JUAN A. ASCACIO-VALDÉS,[1]
RAÚL RODRÍGUEZ-HERRERA,[1] RUTH E. BELMARES,[1]
MIGUEL Á. AGUILAR-GONZÁLEZ,[4] CRISTIAN MARTÍNEZ-ÁVILA,[5]
HELIODORO DE LA GARZA-TOLEDO,[6] and
CRISTÓBAL N. AGUILAR[1]

[1] *Food Research Department, School of Chemistry,
Autonomous University of Coahuila, Boulevard Venustiano
Carranza and José Cárdenas s/n, República Oriente, Saltillo 25280,
Coahuila, México*

[2] *School of Health Sciences UA de C, Calle de la Salud #714,
Villa de Fuente, Piedras Negras, Coahuila, México, C.P. 26090,
E-mail: janethventura@uadec.edu.mx*

[3] *School of Medicine, North Unit UA de C, Calle de la Salud #714,
Villa de Fuente, Piedras Negras, Coahuila, México, C.P. 26090*

[4] *Center for Research and Advanced Studies of National Polytechnic
Institute (CINVESTAV) Saltillo Unit, Av. Industrial Metalúrgica #1062,
Parque Industrial Saltillo-Ramos Arizpe, Ramos Arizpe, Coahuila,
México, C.P. 25900*

[5] *School of Agronomy, Autonomous University of Nuevo Leon,
Francisco Villa S/N, Col. Ex-Hacienda el Canadá, Gral. Escobedo,
Nuevo León, México, C.P. 66050*

[6] *Antonio Narro Agrarian Autonomous University. Blvd. Antonio
Narro 1923 Col. Buenavista, Saltillo, Coahuila, México, C.P. 25315*

## ABSTRACT

The objective of this research is to show the genetic diversity of maize varieties with pigmented grain present in Mesoamerica, mainly in Mexico. It is intended to draw attention to the preservation of the genetic and cultural heritage of Mexico, pigmented maize varieties, which are currently considered nutraceuticals by the presence of anthocyanins in the grain. These phenolic compounds are responsible for the color presented by these crops, which could vary from red to blue. Anthocyanins have beneficial biological activity since they act as antioxidants helping to oxidative stress prevention and to reduce the risk of no communicable diseases. The anthocyanin content in the grains depends on the genotype of the plant, black or purple grains have a higher content than white, pink, and yellow grains. Pigmented maize varieties have the potential for being used in the development of new functional foods as well as for replacing artificial dyes in the food industry. The commercial and industrial potential of these corns is promising; however, they are being threatened by the planting of genetically modified maize (MGM), which could bring serious consequences in its genetic purity and conservation. Therefore, the protection of these corns is urgent, giving them added value to improve their profitability, fact that undoubtedly would impact the economy of farmers who have been key elements in the conservation of these types of corns.

## 4.1 INTRODUCTION

Term "Mesoamerica" was coin by Paul Kirchhoff and refers to the region comprised between Panuco and Sinaloa rivers and the borderline between Guatemala and El Salvador [1]. This region is considered one of the principal places of plant domestication, principally of corns, a milestone of the uprising of different ancient cultures as Olmecas and Teotihuacanos in Mesoamerica and the cultures Inca and Quechua in the Andean region of South America [2]. In the region of Mesoamerica, corns with pigmented grains that have been preserved thanks to the perseverance of the local producers that still raise it for their own consumption. Colored grains can be blue, black, red, purple, etc. [3]. Nowadays, corn production has been focused on raising white grain genotypes, with pigmented corn been underestimated by the politics on genetic improvement and commerce [4]. In this chapter, we will talk about the value added of pigmented grain maize and about the importance of preserve these varieties in the center of origin.

## 4.2  ORIGIN OF CORN AND ITS ROLE IN THE EVOLUTION OF MESOAMERICA

Many scientists have discussed the origin of maize (*Zea mays* L), the cob is unique between cereals and the explanation of its origin is a great scientific challenge [5]. Vavilov and Bukasov mentioned that genetic diversity of culture plants in concentrated in some regions around the world called "center of origin and diversity." Mesoamerica is one of the seven regions originally named as a member of this list by Vavilov [6, 7] being of principal importance between its original cultivars the corn due to its vast genetic diversity and the presence of wild related species and subspecies. The corn played a key role in the ancient civilization of Mesoamerica as shown the chronicles, songs, legends, and mythology [8].

There are many theories about the origin of maize, nevertheless the most accepted place the *teocintle* as a genetic ancestor of this crop [9–11]. Genetic evidence has demonstrated that *teocintle* species Z. *mays* spp. Parviglumis is the direct ancestor of corn [10, 11]. Species of *teocintle* and other crops related to the genre *Tripsacum* ("maicillos") are restricted to tropical and subtropical areas of Mexico, Honduras, Guatemala, and Nicaragua [8].

## 4.3  GENETIC DIVERSITY OF PIGMENTED MAIZE IN MEXICO

Corn is one of the most important cereals in the diet of Mexican people; it represents an invaluable good for cattle raising and is also the principal raw material for the manufacturing of numerous industrial products. In 2014, Mexico produced more than 23 million ton of corn, as depicted in Table 4.1 [12].

The white corn is commonly the most consumed and its production represents the 90% of total national production; by the other hand maize with colored grain means a minimal fraction of the total production (Table 4.1) because is produced only in very specific regions and isn't spread over the whole national territory. An important point to remark is that

**TABLE 4.1**　Detailed Production of Corn in Mexico

| Cultivar | Type /Variety | Production (Ton) |
|---|---|---|
| Corn grain | White | 20,710,883.68 |
| Corn grain | Yellow | 2,422,715.12 |
| Corn grain | Colored | 86,821.90 |
| Corn grain | "Pozolero" | 38,593.26 |
| Corn grain | | 14,242.58 |
| Popcorn grain | | 2,358.14 |
| Total | | 23,275,614.68 |

commercial price of blue corn grain is above 70% over white maize price [13]. By another side, Argentina is the only one country worldwide that exports about 400 ton by year of colored corn grain type Flint to the European Union [14]. From an economic point of view, maize is the principal culture in Mexican territory a means the 18% of the total value of the agricultural activity, generating more than 4800 millions USD in 2012 and more than 4300 millions USD in 2013 and concentrates the 33% of cultures surface of Mexican territory (7.5 million ha) [12].

The big genetic diversity of maize allows the existence of species with a variety of color that goes from yellow, black, purple, blue, red, and orange but not only the mentioned here; presence of these colors is attributable to the high content of phenolic compounds of the flavonoid group, mainly anthocyanins [15]. Recently, the interest of the scientific community over anthocyanins has grown due to its possible benefits to human health such as reduction of coronary sickness, anticancer activity and its antioxidant function [16]. Colored maize has been considered as nutraceuticals and functional foods due to the value added that provides the presence of antioxidant compounds that confer a higher commercial cost compared with white grain corn [13]. The content of anthocyanins in the grain is variable due to the difference in the races and place of culture. Genetic, biochemical, and molecular analysis of anthocyanins synthesis have demonstrated that is necessary an enzymatic conversion from leucoanthocyanins to anthocyanins to develop coloration, this conversion is controlled by gene anthocyaninless 2 (a2) [17]. Mendoza, in 2012 determined that the grain of colored maize exists a pattern of anthocyanins accumulation in the plant and each of its organs. The accumulation sequence starts in the foliar pods, foliar sheets, stalks, ears, young corns, steams, cob and it ends in the grain [18].

The characterization of dyes in the corn grain has been studied for more than 50 years, but the newest reports are the most complete. Since the beginning of the present century, some anthocyanins have been identified in maize foils and flowers, these are cyanidin 3-glucoside; cyanidin 3-(6"-malonylglucoside); cyanidin 3 (3,6 dimalonylglucoside); peonidin 3-glucoside and peonidin 3-(dimalonylglucocyde) [19]. In 2004, Kuskoski et al. demonstrated that cyanidin 3-glucoside have an antioxidant capacity even major than Trolox (a standard reference in these analytical techniques) [20]. The content of cyanidin 3-glucoside in different genotypes of maize is shown in Table 4.2, classified into three levels: low, medium, and high. Since anthocyanins are responsible for the grain coloration, it can be concluded that white corns have negligible concentrations of these compounds, being the yellow waxy corn Xinnuo 301 the one with the lower concentration (0.63± 0.05 mg/100 g of dry grain) and the purple corn variety Zhuozhuo who present the bigger concentration (3045 ± 163.2 mg/100 g of dry grain) [21]. Nevertheless, the absence of color in grains is important to consider the existence of corn of white grain and red cobs-

TABLE 4.2   Anthocyanins Content in Different Genotype of Maize With Pigmented Grain.

| Variety | Anthocyanins content Average ± SD* (mg eq of cyanidin 3 glucoside/100g) | Reference |
|---|---|---|
| Yellow waxy (Xinnuo 301) | 0.63 ± 0.05 | [22] |
| Black pearl⌐ Guangdong (Hybrid). | 29.22 ± 0. 86 | [23] |
| Blue Shaman | 32.27 ± 0.15 | [24] |
| Red Purple sweet | 60.71 ± 2.17 | [24] |
| Red commercial | 82.3 ± 3.8 | [25] |
| Red crimson Shijazhuang | 149.3 ± 5.63 | [23] |
| Black mexican red | 324 ± 3.45 | [26] |
| Mexican mottled (Red) | 431 ± 3.89 | [26] |
| Purple Zhuozhou | 3045 ± 163.2 | [23] |

*SD = Standard deviation.

Adapted from Escalante-Aburto et al., [21]

Some authors report that there are about 56 races of corn that have been identified, many of them present pigmented varieties [9, 11, 27–29].

To understand the difference between a native variety and a creole one, is important to say that native varieties have evolved in a natural way into a specific location and probably lack of the capacity to adapt to a different environment because its development depends on specific geographic and climatic conditions; creole varieties are hybridized naturally or by human intervention, this corns evolved naturally thanks that the native ancestors and its successor take the decision of keeping its culture [30]; then we can say that creole varieties are those that are commonly exploited because the major part of farmers chooses to culture these seeds due to its resistance and its predictability comparing with artificially improved seed [31].

Kuleshov and Vavilov were very impressed with the great diversity of maize in Mexico and Guatemala [7, 32]. Vavilov wrote: "Central America, including the southern part of Mexico, is of a great interest, especially as the center of origin of the corn. Researches in Mexico and Guatemala have revealed a great potential in varieties, particularly in the mountain region of southern Mexico. There is no doubt that morphological diversity of corn varieties, that there is not any similar in another country, is focused on the diversity of physics and ecologic types that have been underused in genetic improvements" [7].

Corn is sowed in practically all the territory of Mexico and is cultivated through arid zones in hot and wet weathers, in these last conditions is where to exist the major genetic diversity of this crop. In the major part of the cases, diversity is bounded to ecological conditions nevertheless, human interventions have played a key role in the creation of that diversity [8].

In the group of colored corns, blue grains are predominant because it presents a consumer preference, principally in the center of the country. In the same way, the center region of the country is where a major diversity has been found due to its geographical characteristics like low environmental temperatures, frosts, and lack of rain between November and April [33]. Serratos-Hernandez et al. [34] estimated that a third part of the producers of the seasonal cultivar of maize in the State of Mexico sows near of a 33% of their lands with blue corn. Programs focused on corn-pigmented grains have been developed in the Center for Research for Improvement of Maize and Wheat (CIMMYT in Spanish) due to its sweeter flavor compared with other varieties raised for human consume.

The granulated consistency of these corns makes a little more dense torti-llas than those made with white corn flour [3]. There are several varieties of blue corn [35, 36], besides creole varieties, some hybrids have been developed for example, in the State of Mexico exists is commercialized a hybrid named "Black Carioca" [36]. The National Institute of Forestall, Agricola and Livestock Researchers (INIFAP in Spanish) have devel-oped the varieties "V–39 Cocotitlán," "V–45 Sierra Negra," "V–43 AZ Malintzi" among others [37, 38]. In poorly technified areas, is possible to find a major diversity of colored grain corns thanks to the differences in environmental conditions, management, and usage [8]. Information about the modern variability of maize in Mexico and its relations with the weather, agronomic, morphological, and isoenzymatic factor [8] are summarized in Table 4.3.

The impact of corn in Mexico is really impressive, its importance is not just alimentary but cultural since through time many mites and believing have been attributed to this crop. For example, there are still some places where ceremonies dedicated to red corn are performed to achieve fertility and a good yield of the harvest [44].

## 4.4 TYPICAL FOOD MADE WITH CORN

In the same way, Mexico is rich in genetic diversity of maize and in corn-based gastronomy. The United Nations Educational, Scientific, and Cultural Organization have considered Mexican food as Intangible Cultural Heritage of mankind [45]. The corn-based gastronomy is a part of the collective memory and not only comprehends food intake besides express social and economic relationships and make patent acts deeply charged of symbolisms [46].

Culinary antecedents of corn date from Medium Preclassic Age (1200–400 B.C.) [47]. According to data from the Codexes Florentino and Mendocino, it is possible to conclude that maize was the predomi-nant component in the diet of Mesoamerica; although there were other plants [48]. Some probable uses for corn in prehispanic age could be flour making, Pinole, and heat-popped grains (popcorns) [49]. These prepara-tions are made with native corns because the artificially improved grains do not have the necessary properties to be a suitable raw material for the dishes [50]. Some typical foods that can be prepared with any white

**TABLE 4.3**    Different Races of Pigmented Maize in Mexico (Images from CONABIO, [29])

| Samples of colored maize | Brief description |
|---|---|
| **Races from higher land in the center region of Mexico**  Sample of race corn Sweet. | Denominated conical group, includes Northern Conical, "Palomero Toluqueño," Conical, Conical Corn, "Arrocillo," "Chalqueño," "Mushito," "Cacahuacintle" and Sweet Corn. This group has in common conical cobs, reduced number of leafs in the ear, weak roots system, fallen leafs and strongly pubescent ear of leafs with the presence of anthocyanins [8]. |
| **Races from higher land in the northern region of Mexico**  Sample of race "Apachito." | It is possible to find here the group "Sierra de Chihuahua," that includes "Cristalino de Chihuahua," "Gordo," "Azul," "Apachito" and "Complejo Serrano de Jalisco" [39]. These races are found in the high lands of northeastern Mexico (mainly Chihuahua and some regions of Sonora, Durango, and Jalisco) in little valleys about 2000 and 2600 high meters [27, 39–41]. |
| **Races from higher land in the occidental region of Mexico (group of eight lines of grain)**  Sample of "Elotes Occidentales." | This group includes races distributed in low elevations in western and north of Mexico like "Harinoso de Ocho," "Tabloncillo Perla," "Tabloncillo," "Bofo," "Elotes Occidentales," "Blando de Sonora" and "Onaveño" [29]. |
| **"Chapalote"** 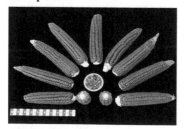 Sample of "Elotero de Sinaloa." | The race Chapalote was described as one of the most distinctive from Mexico [41] and as one of the most ancient of this country [42]. Races of this group include "Chapalote," "Reventador," "Dulcillo del Noroeste" and "Elotero de Sinaloa." These corns have a brown pericarp and crystalline grains; "Dulcillo del Noroeste" have sweet endosperm and very variated colors in grain; "Elotero de Sinaloa" have mixed grain colors; "Reventador" have white and crystalline grains [8, 29]. |

**TABLE 4.3** *(Continued)*

| Samples of colored maize | Brief description |
| --- | --- |
| **Races of tropical corns of early maturation**  Samples of race "Zapalote Chico." | This group includes races of short cycles and low sensibility to photoperiod adapted to low elevations and distributed mainly in coastal plains of the Pacific and the Peninsula of Yucatan. Races in this group are "Conejo," "Zapalote Chico," "Nal-Tel" and "Ratón," this last one is distributed in the northern region of Mexico in low and middle elevation (100–1300 m). [29, 43]. |
| **Group of late maturity of southeaster of Mexico**  Samples of race "Motozinteco." | This group includes races sowed in middle altitudes in the south of Mexico and is characterized by very late plants (95 to 115 days of floration), with high sensitivity to photoperiod and temperature. Races in this group are "Olotillo," "Dzit-Bacal," "Comiteco," "Motozinteco," "Tehua," "Olotón" y "Coscomatepec" [8, 29, 41]. |

or colored corn are tortillas, tortilla chip, cooked corn, soups, tamales, atole, and other beverages [50, 51]. Pigmented maize can also be used in the elaboration of traditional and fermented beverages [52]. In Mexico, different types of maize have active markets, like blue corn (to making foods), red or pink (to feedstock or making tortillas) (Figure 4.1) [3]. Nowadays many alternatives of alimentary products crafted with colored maize have appeared, as an example, it is possible to find pizza of blue corn dough and beer of fermented red corn grains.

**FIGURE 4.1**    Red corn tortilla.

Several studies have evaluated changes in content or profile of antho-cyanins in food products crafted by "nixtamalization," an ancient process transmitted from generation to generation in Mesoamerica, that consist in the addition of about 1% of lime to the corn cooking process [53]. This technique is used in the manufacturing of a great variety of products (torti-llas, atole, snacks among other) and involves changes in chemical, struc-tural, and nutritional nature in many compounds of the grains [54–57]. Traditional nixtamalization is an aggressive process for anthocyanins and can provoke even a total loss of these compounds in the products obtained from colored corn compared with the content in the grains [21]. The tradi-tional process has kept identically to our days and it used by the producers dedicated to the commerce of typical snacks. The foods obtained from colored maize have gained a raising interest and a better market posi-tion in the food industry and commerce thanks to its content of bioactive compounds.

## 4.5  APPLICATIONS OF ANTHOCYANINS FROM MAIZE

Some recent investigations have been reviewed the applications of the anthocyanins obtained from pigmented maize (Table 4.4) in different areas.

**TABLE 4.4** Uses and Applications of Anthocyanins from Colored Maize

| Research | Results | Reference |
| --- | --- | --- |
| Antioxidant activity of methanolic and aqueous extracts from several varieties of Mexican corn. | Varieties "red," "orange," "blue," and "RO" (Red Olinala) hold 50–82% of water-extractable phenolic compounds relative to methanol; "white" variety exhibited the lower presence of these compounds. | [26] |
| Importance of market niche. A case study of blue corn and pozole corn in Mexico. | Blue corn and "pozolero" corn represent a big challenge for specialized markets, it means that both grains itself have potential to reduce poverty, it is difficult to reach if only a reduced group of farmers and processor relatively more richest gets the major part of the profits. | [30] |
| Nixtamalization and its effect on the content of anthocyanins of pigmented maizes. | The new options of nixtamalization, like fractioned and extrusion, have shown that is possible to retain higher quantities of anthocyanins (38–58%) in corn products made with these technologies. | [21] |
| Development of biodegradable films based on blue corn flour with potential applications in food packaging. Effects of plasticizers on mechanical, thermal, and microstructural properties of flour films | Blue corn flour seems to be an interesting source of raw material for the production of biodegradable films. Films made with flour and sorbitol as plasticizer shown an improved adhesion and a compacted structure. | [58] |
| Anthocyanins and antioxidant activity in maize (Zea mays L) of races "chalqueño," "conic corn" and "bolita." | These races present the same anthocyanins and the profile was characterized by the predominance of acylated anthocyanins (>60%) derived from cyanidin, the most abundant aglycone. | [36] |
| Quality attributes and anthocyanin content of rice coated by purple-corn cob extract as affected by coating conditions | Results shown that anthocyanins are susceptible to thermic treatment. To obtain a major quantity and biological activity of rice covered with anthocyanins of purple corn is recommended to use cover conditions with low temperatures and longer time of pulverization. | [59] |
| Antioxidative effect of purple corn extracts during storage of mayonnaise | Extract from purple corn leaf is rich in anthocyanins and has a good antioxidant activity in mayonnaise and is able to use it in mayonnaise products. Natural antioxidants are used to raise the shell life of food products and to give protection to mayonnaise substituting synthetic antioxidants as BHT and EDTA. | [60] |

**TABLE 4.4** *(Continued)*

| Research | Results | Reference |
|---|---|---|
| Optimization of microwave-assisted extraction of anthocyanins from purple corn (*Zea mays* L.) cob and identification with HPLC–MS | Microwave-assisted extraction (MAE) is more efficient than conventional techniques. Total Anthocyanins Content was incremented depending directly on extraction time. | [61] |
| Effect of supplementation of purple pigment from anthocyanin-rich corn (*Zea mays* L.) on blood antioxidant activity and oxidation resistance in sheep | Ingest of anthocyanins from purple corn rise Superoxide dismutase enzymatic activity and strengthen antioxidant activity and plasma from sheep with no oxidative stress. | [62] |
| Phytochemicals and Antioxidant Capacity of Tortillas Obtained after Lime-Cooking Extrusion Process of Whole Pigmented Mexican Maize | Tortilla made with pigmented Mexican corn flour retained 76.4–87.5% of total phenols, 27.1–65.4% of anthocyanins and 87.2–90.7% of antioxidant capacity but there is necessary more research to determine the implication of these types of maize in Mexican people health. | [63] |
| Identification and antioxidant activity of anthocyanins extracted from the seed and cob of purple corn (*Zea mays* L.) | Cyanidin–3-glycoside, pelargonydin–3-glucoside, and peonidina–3-glycoside are components of extracts from purple corncob. | [64] |
| Nutraceutical profiles of improved blue maize (*Zea mays*) hybrids for subtropical regions | Hybrids with high levels of nutraceuticals compared to local varieties of blue corn were found. These hybrids are an option for commercial production of tortillas and related food products. | [65] |
| Acetylation of normal and waxy maize starches as encapsulating agents for maize anthocyanins microencapsulation | Acetylated starches present a better behavior as encapsulating agents for anthocyanins from purple corn, even better than hydrolyzed starches. | [66] |
| Effect of traditional nixtamalization on anthocyanin content and profile in Mexican blue maize (*Zea mays* L.) landraces | Raw blue corn presented an anthocyanins profile with high presence of cyanidins derivatives in an average of 86.9%, while nixtamalization incremented the relative presence of glycosylated anthocyanins (CY–3-Glu, y PG–3-Glu) and reduced acylated anthocyanins (Cy-Suc-Glu, y Cy-diSuc-Glu) compared to raw crop. | [67] |

**TABLE 4.4**  *(Continued)*

| Research | Results | Reference |
|---|---|---|
| Effects of baking conditions and dough formulations on phenolic compound stability, antioxidant capacity, and color of cookies made from anthocyanin-rich corn flour | Results shown that citric acid raises significantly the content of total flavonoids and anthocyanins in cookies prepared from dark red and blue corns. | [68] |
| Dye-sensitized solar cells based on purple corn sensitizers | Extracts from purple corn byproducts (peel, corn, and stigma) can be potentially used as photosensitizers in dye-sensitized solar cells (DSSC) with the aim of developing a high-performance, low-cost, and environmental-friendly DSSC. | [69] |
| *In vitro* digestibility, crystallinity, rheological, thermal, particle size and morphological characteristics of Pinole, a traditional energy food obtained from toasted ground maize | Analysis shown that toast cycle induced important effects in morphology and digestibility of white and blue corn flour. | [70] |
| Multiple optimizations of chemical and textural properties of roasted expanded purple maize using response surface methodology | Optimized product presents good nutritional characteristics and texture. The Intensification of Vaporization by Decompression to the Vacuum became an interesting process of texturing of purple corn that may preserve its antioxidants properties. | [71] |

Information in Table 4 show that different countries are interested in obtaining anthocyanins from corn grains to apply them in the food area as natural dyes or antioxidant; the ranching industry present also interest in these compounds due its phenolic nature, the principal use is as antioxidants is for improving health and prevent the oxidative stress if the cells but at this moment there are none or few reports that support its effectiveness and bioavailability in human beings.

## 4.6   POTENTIAL RISK FOR CORN OF PIGMENTED GRAINS

Between the potential risks that can affect the conservation of the purity of these crops, one of the worst menaces is the genetic contamination by genetically modified corns (GMC). Since an economic view, MGC's are patented and belongs to particular enterprises and the seeds cannot be traded or spread without the corresponding property title becoming the privatization of these public properties a potential risk [72]. As we stated before maize in Mexico represents a wealth not just in biodiversity but additionally economic, agricultural, social, and cultural, not as in the United States of America (USA) that is considered as the leader in the genetically modified organism (GMO). Is estimated that more than two million of families sow 59 natives races of corn in approximately 8 million Ha (66% of total national sowed surface) [73]. Although farming of corn is a common practice in the USA, it is not classified as transgenic and non-transgenic corns, becoming the importations to Mexico into a risk of contamination of the genome of native corns with MGMs.

Different research centers have declared be worried about the farming and commercialization of MGM cultivars, principally by the threatening that means to the natural genetic diversity (INIFAP, Colegio de Postgraduados en Ciencias Agrícolas, Universidad Autónoma de Chapingo, Universidad Autónoma Agraria Antonio Narro and Universidad Nacional Autónoma de México). The unconformity of these institutions makes reference to the apparition of unfair practices in the sector of seeds production after the vanish of the National Seed Producer (Productora Nacional de Semillas in Spanish, PRONASE), this event originates the opening of the sector to multinational enterprises that commercialize OMGs at high prices [74]. However, these institutions have promoted the use of native and improved seeds produced in the country.

On March 18th of 2005, the Law of Biosecurity on Genetically Modified Organisms (LBGMO) was published in the Official Diary of the Federation of the Nation. This law has the objective of "make a regulation of the activities of confined use, experimental liberation, pilot program liberation, commercial liberation, commercial importation and exportation of GMO to prevent, evading or reduce the possible threat that these activities may mean to human health, environment, biological diversity or animal, vegetal, and aquatic health [75]. Nevertheless, this law has not succeeded completely. There have been detected the presence of transgenic corns in several national cultivars [76]. This is due principally to the importation of GMO's from the USA. In front of the imminent entrance of GMO's to Mexico, is convenient to inform to the producers that improved national seeds have counterparts equivalent or even superior to imported GMO's seeds and present a lower price (till 50% cheaper) [77]. Another critical point in the conservation of pigmented corn genotype is the farmers now that they select the seed to sow. The CIMMYT mentioned that if little farmers had access to transgenic corn varieties and they consider they are most valuables than native seeds, they could promote its diffusion in the local corn populations, leading to a genetic contamination of native corns affecting the genome of this races, although is hard to determine the affectation degree [12]. The reality is in the parcel, where the family farmers face the challenges of rain seasons and select the more convenient seed according to the environmental conditions, here is where the uncertainty of GMC's seeds is generated.

A variety of creole maize is due in a big proportion to the introduction of improves seeds and the increase in the migration of rural populations [78]. Sarmiento and Castañeda [79] mentioned that since the race diversity does exist, laws and public politics about biosecurity, through LBGMO to the Regimen of Special Protection of Corn and the Program for the Conservation of the Creole Maize (elaborated by the National Commission for the Protected Natural Areas, CONANP), have disposition only in a general view that cannot aboard the local problems. The generality and its limitation in its declarations contribute scarcely to the local preservation of the native corn varieties.

## 4.7   FINAL COMMENTS

The contribution that Mesoamerica has given to the world on the theme of genetic diversity of maize is evident. Mexico is the richest country in corn varieties and has a special mention in the variety of colored maize. In this and other countries of Mesoamerica is important to keep the genetic compendium of this crop. Colored corns are generally of a low-profile commercialization and used originally in important social and religious events but nowadays have gained a major projection in the general commerce, including international business. Government politics does not give enough protection to farmers in the aim to ensure their production and save the importance of their cultivars to the economy and the culture in a familiar and national context. Regarding the risk that GMC represents, is important to not exaggerate since, finally, GMO's are biotechnological tools useful for certain objectives; however, they do not must be liberated to commercial sow in the centers of origin of corn under no circumstances due to the risk of genetic contamination and the privatization of a public wealth. In the same way, the promotion of the native seed must be strengthened and the rural development supported somehow. Participation of institutions like INIFAP and Universities; and No Governments Organizations is very important to the preservation of this natural resource. However, the same must happen with the other Mesoamerican countries also riches in corn diversity. Researchers with pigmented corns are related to its content of anthocyanins due to its antioxidant effect. Until this day, there are no documents about its effectiveness in humans.

## KEYWORDS

- **anthocyanins**
- **pigmented maize**
- **preservation**

## REFERENCES

1. Kirchhoff, P., (1960). Mesoamerica its geographical limits, ethnic composition and cultural characteristics. *Supl. la Rev. TLATOANI1, 3,* 13.
2. Carrillo, T. C., (2009). The origin of corn nature and culture in Mesoamerica. *Ciencias.,* pp. 4–13.
3. Hellin, J., & Keleman, A., (2013). The Creole Corn Varieties, the Specialized Markets and the Life Strategies of the Producers. *LEISA Revista de Agroecologia, 29*(2), 9–14

4. Muñoz, O. A., (2005). Deciphering The diversity of corn from the ecological niches of Mexico. In: Centli-corn: Prehistory and History, Diversity, Potential, Genetic and Geographic Origin. *Postgraduate School.* ISBN 968-839-458-0.
5. Serratos, H. J., (2009). In The Origin and Diversity of Corn in the American Continent. *Green Peace* Mexico. Mexico City.
6. Bukasov, S. M., (1930). In *The Cultivated Plants of Mexico, Guatemala, and Colombia.* Turrialba, Costa Rica.
7. Vavilov, N. I., (1931). *Mexico and Central America as the Principal Centre of Origin of Cultivated Plants of New World.* Bulletin 1, Leningard: *Bulletin of Applied Botany and Plant Breding.*
8. Sánchez, G. J., (2011). *Diversity of corn y teosinte.* Report for the project collection, generation, actualization and analysis of information about genetic diversity from corn and its wild relatives in Mexico. National Commission for the Knowledge and Use of Biodiversity. Guadalajara, Mexico.
9. Beadle, G., (1939). Teosinte and the Origin of Maize. *J. Hered., 30,* 245–247.
10. Doebley, J. F., (1990). Molecular evidence and the evolution of maize. *Econ. Bot., 44*(1), 6–27.
11. Matsuoka, Y., Vigouroux, Y., Goodman, M. M., Sanchez, G, J., Buckler, E., & Doebley, J., (2002). A single domestication for maize shown by multilocus microsatellite genotyping. *Proc. Natl. Acad. Sci. USA, 99*(9), 6080–6084.
12. Cruz, M. S., Gómez, M. M., Entzana, A. M., Suárez, C. Y., & Santillán, V., (2012). Report Actual situation and perspectives from corn in Mexico *1996–2012,* SIAP: Mexico City.
13. Arellano, V. J. L., Rojas, M. I., & Gutiérrez, H. G. F., (2013). Hybrids and synthetic varieties of blue corn for the central highlands of Mexico: Agronomic potential and yield stability *Rev. Mex. Ciencias Agrícolas, 4*(7), 999–1011.
14. Alvarez, A., (2006). Application of corn in technology and other industries. *Inf. Espec. ILSI Argentina Vol. II,* 10–11.
15. Masuoka, N., Matsuda, M., & Kubo, I., (2012). Characterization of the antioxidant activity of flavonoids. *Food Chem., 131*(2), 541–545.
16. Norberto, S., Silva, S., Meireles, M., Faria, A., Pintado, M., & Calhau, C., (2013). Blueberry anthocyanins in health promotion: A metabolic overview. *Journal of Functional Foods,* pp. 1518–1528.
17. Cone, K. C., (2007). Anthocyanin synthesis in maize aleuronic tissue. *Plant Cell Monogr., 8,* 121–139.
18. Mendoza, M. C. G., (2012). Corn anthocyanins: Its distribution in the plant and production. Corn anthocyanins: Its distribution in the plant and production. Colegio de Posgraduados.
19. Fossen, T., Slimestad, R., & Andersen, O. M., (2001). Anthocyanins from maize (*Zea mays*) and reed canarygrass (*Phalaris Arundinacea*). *J. Agric. Food Chem., 49*(5), 2318–2321.
20. Kuskoski, E. M., Asuero, A. G., García-Parilla, M. C., Troncoso, A. M., & Fett, R., (2004). Antioxidant activity of anthocyanin pigments. *Ciência e Tecnol. Aliment., 24*(4), 691–693.

21. Escalante-Aburto, A., Ramírez-Wong, B., Torres-Chávez, P. I., Barrón-Hoyos, J. M., Figueroa-Cárdenas, J. De D., & López-Cervantes, J., (2013). The nixtamalization and its effect on the anthocyanin content of pigmented maize, a review. *Rev. Fitotec. Mex.*, *36*(4), 429–437.

22. Hu, Q. P., & Xu, J. G., (2011). Profiles of carotenoids, anthocyanins, phenolics, and antioxidant activity of selected color waxy corn grains during maturation. *J. Agric. Food Chem.*, *59*(5), 2026–2033.

23. Zhao, X., Corrales, M., Zhang, C., Hu, X., Ma, Y., & Tauscher, B., (2008). Composition and thermal stability of anthocyanins from Chinese purple corn (*Zea mays* L.). *J. Agric. Food Chem.*, *56*(22), 10761–10766.

24. Abdel-Aal, E. S. M., Young, J. C., & Rabalski, I., (2006). Anthocyanin composition in black, blue, pink, purple, and red cereal grains. *J. Agric. Food Chem.*, *54*(13), 4696–4704.

25. Lopez-Martinez, L. X., Parkin, K. L., & Garcia, H. S., (2011). Phase II-inducing, polyphenols content and antioxidant capacity of corn (*Zea mays* L.) from phenotypes of white, blue, red and purple colors processed into masa and tortillas. *Plant Foods Hum. Nutr.*, *66*(1), 41–47.

26. López-Martínez, L. X., & García-Galindo, H. S., (2010). Antioxidant activity of methanolic extracts and aqueous from different varieties of Mexican corn. *Nov. Sci.*, *2*(3), 51–65.

27. Ortega, P. R. A., (1985). Mexican variates and breed form maize and its evaluation in crossing with temperature lines as initial material for plant breeding. Ph D. Thesis. N. I. Vavilov National Institute of plants. Leningard, U.R.S.S.

28. Piperno, D. R., & Flannery, K. V., (2001). The earliest archaeological maize (*Zea mays* L.) from highland Mexico: New accelerator mass spectrometry dates and their implications. *Proc. Natl. Acad. Sci. USA*, *98*(4), 2101–2103.

29. Lazos, E., & Chauvet, E., (2012). Analysis of social and biocultural context from several native Mexican corn collected. *CONABIO*: México, City p. 498.

30. Hellin, J., Keleman, A., López, D., Donnet, L., & Flores, D., (2013). The importance of the market niches, a case study for blue corn and corn for pozole in Mexico. *Rev. Fitotecnia Mex.*, *36*, (315–328).

31. Hernández, A. A., & Jordán, C. A., (2001). Why improved maize (*Zea Mays*) varieties are utopias in the highlands of central Mexico. *Converg. Rev. Ciencias Soc.*, *8*(25).

32. Kuleshov, N. N., (1933). World's diversity of phenotypes of maize. *Agron. J.*, *25*(10), 688–700.

33. Eagles, H. A., & Lothrop, J. E., (1994). Highland maize from central Mexico – Its origin, characteristics, and use in breeding programs. *Crop Science*, pp. 11–19.

34. Serratos-Hernandez, J. A., Islas-Gutierrez, F., Buendía-Rodríguez, E., & Berthaud, J., (2005). Gene flow scenarios with transgenic maize in Mexico. *Environ. Biosaf.*, *3*, 149–157.

35. Agama-Acevedo, E., Salinas-Moreno, Y., Pacheco-Vargas, G., & Bello-Pérez, L. A., (2011). Chemical and Physical characteristics of two race from blue corn: starch morphology. *Rev. Mex. Ciencias Agrícolas*, *2*(3), 317–329.

36. Salinas-Moreno, Y., Pérez-Alonso, J. J., Vázquez-Carrillo, G., Aragón-Cuevas, F., & Velázquez-Cardelas, G. A., (2012). Anthocyanins and antioxidants activity from corn (*Zea mays* L) in the races chalqueño, conicos and bolita. *Agrociencia, 46*(7), 693–706.

37. INIFAP, (2006). Technological innovations 2004 (To improve the competitiveness and sustainability of la competitividad y sostenibilidad of agrifood and agroindustrial chains) Brochure; México D. F., *1004.*

38. INIFAP, (2011). *Annual Activity Report 2010*, First National Institute of forestal, agricultural and livestock research: México.

39. Sanchez, J. J., Stuber, C. W., Goodman, M. M., & Sanchez, G. J. J., (2000). Isozymatic and morphological diversity in the races of maize of Mexico. *Econ. Bot., 54*(1931), 43–59.

40. Hernández, E., & Flores, G. A., (1970). Morphologic study of five new races of maize from Sierra Madres Occidental: phylogenetic and phytogeographic implications. *Agrociencia Chapingo*. 749–750.

41. Wellhausen, E. J., & Mangelsdorf, P. C. R., (1951). Races from corn in Mexico: its origin, characteristics and distribution. SAGARPA and Rockefeller note. Mexico.

42. Mangelsdorf, P. C., (1975). Corn; Its origin, evolution and improvement. *Exp. Agric., 11*(4), 326.

43. Sánchez, G. J. J., (2011). Diversity of corn and teosinte. Manus, 2.

44. Vargas, P. E., (2010). The legitimation of royalty among the Maya of the late Preclassic, the mask of the tiger, campeche. *Estud. Cult. Maya, XXXVI*, 13–35.

45. UNESCO. Traditional Mexican cuisine – ancestral, ongoing community culture, the Michoacán paradigm – intangible heritage – Culture Sector – UNESCO https://ich.unesco.org/en/RL/traditional-mexican-cuisine-ancestral-ongoing-community-culture-the-michoacan-paradigm-00400 (accessed Jul 21, 2017).

46. García Urigüen, P., (2012). The food of Mexicans. *Inf CANACINTRA*. México.

47. Taube, K. A., (1989). The maize tamale in classic Maya diet, epigraphy, and art. *Am. Antiq., 54*(1), 31–51.

48. Ortega-Paczka, R., (2003). The maize diversity in Mexico. Esteva, G., & Marielle, C., (eds.). Without maize there is no country. *Inf National Council for Culture and Arts*, 123–154. México.

49. Mera-Ovando, L. M., & Mapes-Sánchez, C., (2009). In: Kato, Y., & Ángel, T., (eds.). Origin and diversification from corn: an analytical review. *Note of Biology Institute of University National Autonomous of Mexico*. Mexico City.

50. Fernández-Suárez, R., Morales-Chávez, L. A., & Gálvez-Mariscal, A., (2013). Importance of Mexican maize landraces in the national diet: An essential review. *Rev. Fitotec. Mex., 36*(3–A), 275–283.

51. Echeverría, M. E., (2000). In: Echeverría, M. E., & Arroyo, L. E., (eds.). Recetary of maize. *CONACULTA*. Mexico.

52. Jiménez Pérez, C., (2014). Utilization of pigmented maize for the elaboration of a functional food. Thesis from Metropolitan Autonomus University. Mexico.

53. Paredes López, O., Guevara, L. F., & Arturo, B. P. L., (2008). The nixtamalization and the nutritive value from corn. *Ciencias, Octubre*, 60–70.

54. Serna-Saldívar, S. O., Knabe, D. A., Rooney, L. W., Tanksley, Jr, T. D., Rooney, D. A., & Tanksley, Jr, T. D., (1987). Effects of lime cooking on energy and protein digestibilities of maize and sorghum. *Cereal Chem., 64*(4), 247–252.
55. Gomez, M. H., Mcdonough, C. M., Rooney, L. W., & Waniska, R. D., (1989). Changes in corn and sorghum during nixtamalization and tortilla baking. *J. Food Sci., 54*(2), 330–336.
56. Bressani, R., (1990). Chemistry, technology, and nutritive value of maize tortillas. *Food Rev. Int., 6*(2), 225–264.
57. Rojas-Molina, I., Gutierrez-Cortez, E., Palacios-Fonseca, A., Baños, L., Pons-Hernandez, J. L., Guzmán-Maldonado, S. H., et al., (2007). Study of structural and thermal changes in endosperm of quality protein maize during traditional nixtamalization process. *Cereal Chem., 84*(4), 304–312.
58. Valderrama, S. A. C., & Rojas, G. C., (2014). Development of biodegradable films based on blue corn flour with potential applications in food packaging. Effects of plasticizers on mechanical, thermal, and microstructural properties of flour films. *J. Cereal Sci., 60*(1), 60–66.
59. Duangkhamchan, W., & Siriamornpun, S., (2015). Quality attributes and anthocyanin content of rice coated by purple-corn cob extract as affected by coating conditions. *Food Bioprod. Process, 96*, 171–179.
60. Li, C. Y., Kim, H. W., Li, H., Lee, D. C., & Rhee, H. I., (2014). Antioxidative effect of purple corn extracts during storage of mayonnaise. *Food Chem., 152*, 592–596.
61. Yang, Z., & Zhai, W., (2010). Optimization of microwave-assisted extraction of anthocyanins from purple corn (*Zea mays* L.) cob and identification with HPLC–MS. *Innov. Food Sci. Emerg. Technol., 11*(3), 470–476.
62. Hosoda, K., Miyaji, M., Matsuyama, H., Haga, S., Ishizaki, H., & Nonaka, K., (2012). Effect of supplementation of purple pigment from anthocyanin-rich corn (*Zea mays* L.) on blood antioxidant activity and oxidation resistance in sheep. *Livest. Sci., 145*(1–3), 266–270.
63. Aguayo-Rojas, J., Mora-Rochín, S., Cuevas-Rodríguez, E. O., Serna-Saldivar, S. O., Gutierrez-Uribe, J. A., Reyes-Moreno, C., et al., (2012). Phytochemicals and antioxidant capacity of tortillas obtained after lime-cooking extrusion process of whole pigmented Mexican maize. *Plant Foods Hum. Nutr., 67*(2), 178–185.
64. Yang, Z., & Zhai, W., (2009). Identification and antioxidant activity of anthocyanins extracted from the seed and cob of purple corn (*Zea mays* L.). *Innov. Food Sci. Emerg. Technol., 11*(1), 169–176.
65. Urias-Peraldi, M., Gutirrez-Uribe, J. A., Preciado-Ortiz, R. E., Cruz-Morales, A. S., Serna-Saldivar, S. O., & Garcia-Lara, S., (2013). Nutraceutical profiles of improved blue maize (*Zea mays*) hybrids for subtropical regions. *F. Crop. Res., 141*, 69–76.
66. García-tejeda, Y. V., Salinas-moreno, Y., & Martínez-bustos, F., (2014). Food and bioproducts Processing acetylation of normal and waxy maize starches as encapsulating agents for maize anthocyanins microencapsulation. *Food Bioprod. Process, 94*, 717–726.
67. Mora-Rochín, S., Gaxiola-Cuevas, N., Gutiérrez-Uribe, J. A., Milán-Carrillo, J., Milán-Noris, E. M., Reyes-Moreno, C., et al., (2016). Effect of traditional nixtamaliza-

tion on anthocyanin content and profile in mexican blue maize (*Zea mays* L.) Landraces. *LWT – Food Sci. Technol.*, *68*, 563–569.

68. Žilić, S., Kocadağlı, T., Vančetović, J., & Gökmen, V., (2016). Effects of baking conditions and dough formulations on phenolic compound stability, antioxidant capacity and color of cookies made from anthocyanin-rich corn flour. *LWT – Food Sci. Technol.*, *65*, 597–603.

69. Phinjaturus, K., Maiaugree, W., Suriharn, B., Pimanpaeng, S., Amornkitbamrung, V., & Swatsitang, E., (2016). Dye-sensitized solar cells based on purple corn sensitizers. *Appl. Surf. Sci.*, 1–7.

70. Carrera, Y., Utrilla-Coello, R., Bello-Prez, A., Alvarez-Ramirez, J., & Vernon-Carter, E. J., (2015). *In vitro* digestibility, crystallinity, rheological, thermal, particle size and morphological characteristics of pinole, a traditional energy food obtained from toasted ground maize. *Carbohydr. Polym.*, *123*, 246–255.

71. Mrad, R., Debs, E., Saliba, R., Maroun, R. G., & Louka, N., (2014). Multiple optimization of chemical and textural properties of roasted expanded purple maize using response surface methodology. *J. Cereal Sci.*, *60*(2), 397–405.

72. Fischer, K., Ekener-Petersen, E., Rydhmer, L., & Björnberg, K., (2015). Social impacts of GM crops in agriculture: A systematic literature review. *Sustainability*, *7*(7), 8598–8620.

73. Perales, H., & Golicher, D., (2014). Mapping the diversity of maize races in Mexico. *PLoS One*, *9*(12).

74. Espinosa-Calderón, A., Turrent-Fernández, A., Tadeo-Robledo, M., San Vicente-Tello, A., Gómez-Montiel, N., Valdivia-Bernal, R., et al., (2014). Law on seed and Federal law on varieties of vegetables and transgenics corns in Mexico. *Rev. Mex. Ciencias Agrícolas*, *5*(2), 293–308.

75. Official Diary of the Federation. (2005) Law on biosafety of genetically modified organism. Mexico organism.

76. Martínez, M. C. R., (2009). Corresponding actions regarding biosecurity of genetically modified organism in the detection of contamination of native corn by transgenic. *Gac. Parlam. Cámara Diput.*

77. Márquez-Sánchez, F., (2008). Summary of the genetic varieties form creole corn (*Zea mays* L) to transgenics hybrids I: germplasm recollections and improved varieties *Agric. Soc. y Desarro.*, *5*(2), 151–166.

78. Escobar, M. D. A., (2006). Peasant valorization of maize diversity a case study of two indigenous communities in Oaxaca, Mexico. Ph D. Thesis, University Autonomous of Barcelone.

79. Sarmiento, B., & Castañeda, Y., (2011). Publics policies aimed at the preservation of native maize varieties in Mexico in the face of agricultural biotechnology the case of Cacahuazintle corn.

# CHAPTER 5

# BIOTECHNOLOGY IMPORTANCE OF POMEGRANATE (*Punica granatum* L.) AND THE USE OF THE PEEL AS AN AGRO-INDUSTRIAL BYPRODUCT

RENÉ DÍAZ-HERRERA,[1]
JANETH MARGARITA VENTURA SOBREVILLA,[2]
and CRISTÓBAL N. AGUILAR[1]

[1] *Group of Bioprocesses, Food Research Department School of Chemistry, Autonomous University of Coahuila, Saltillo, Coahuila, CP 25280, México, E-mail: Cristóbal.aguilar@uadec.edu.mx*

[2] *School of Health Sciences, Autonomous University of Coahuila, Piedras Negras, Coahuila, CP 26090, México*

## ABSTRACT

Food industries that process pomegranate products such as juices, jams, and wines generate large amounts of daily peel (near 50%) producing high contamination due to the incorrect treatment they receive, without giving them a use and only treat them as garbage; among them is the pomegranate peel, a material rich in phenolic compounds with a huge application. This review will discuss about pomegranate (*Punica granatum* L.), its uses, commercial, and medical applications and the potential of this plant at the scientific level, as well as pomegranate peel and the compounds present in it with a seen in the problems related to the extraction and obtaining of these phytochemicals, taking as an alternative the solid-state culture (SSC) technique.

## 5.1  INTRODUCTION

Pomegranate (*Punica Granatum* L), also known as the fruit of Eden in some countries thanks to its flavor and health benefits [1], is a fruit worshiped by many cultures and religions around the world, including Greek Culture, Buddhist, Chinese, Islamic, Hindu, among others. The name "pomegranate" has its origin from the Latin words (*ponus* and *granatus*), which literally translates to "apple with granules." This fruit is native to the Middle East, in countries like Afghanistan, Iran, Pakistan, China, and northern India, where it has been cultivated for thousands of years and is currently cultivated around the world in tropical and subtropical areas such as Egypt, Italy, Chile, Spain, and Mexico, as it requires high temperatures to ripen properly and this fruit can tolerate and grow under drought conditions to temperatures of $-15°C$ [2–8].

*Punica granatum* belongs to the family *Punicaceae* which includes a single genus (*Punica*) and two species (*granatum* and *protopunica*) [3, 9]. Pomegranate is a plant the size of a shrub, which has been reported that extracts from different parts (leaves, bark, flowers, peel, seeds, and fruit) have therapeutic properties (antitumor, antibacterial, antidiarrheal, antifungal, antiulcer) [10, 11].

The world production of pomegranate reaches amounts of approximately 1,500,000 tons, with Iran being the country that contributes the most with 47% of the total production. Iran's pomegranate export has increased almost 95% since 2003 to 2007 (14,075–27,439 tons per year), Unfortunately, about one-third of the food produced for human consumption is lost throughout the processes of production, transportation, storage, etc. Generating amounts of up to 1.3 billion tons of waste per year, Garbage accumulation is a problem all over the world. In places like Europe and North America food waste per capita is between 280–300 kg/year [12], and in countries like South Africa, the accumulation problem is worse, because the range is found at a value of 177 kg/year [13].

This production of pomegranate proves the existing demand for this fruit and its products containing bioactive molecules, which consumers have begun to consume in greater quantity thanks to the benefits of the intake of these products containing polyphenols, due to its antioxidant activity. These phytochemicals represent the largest molecules present in the pomegranate, one of the predominant compounds are hydrolyzable tannins and among them the ellagitannins, especially in peel and mesocarp

[4]. In recent years, due to high consumers demand, progress has been made in research for the purification and extraction of natural compounds for the creation of high-quality products, since the production of bioactive compounds comes mainly from byproducts of pomegranate that are considered as waste [14].

## 5.2  MAIN USES OF POMEGRANATE

The interest in nutraceuticals grows day by day worldwide, thanks to the health benefits they provide. Many studies have been done to demonstrate the beneficial effects of different sources such as fruits, vegetables, and spices, including pomegranate [2]. The first reports on the use of plants for medicinal use come from Mesopotamia in the year 2600 B.C. Egyptian, Chinese, and Hindu medicine also showed knowledge about the application of plant extracts for the treatment of diseases, taking advantage of knowledge of cultures such as Greek and Roman [15]. These products of natural origin have been used for many years for the care and improvement of human health due to the potential of these extracts, so they have been used until now, some as food supplements or even as active ingredients in medicines [2].

Pomegranate is a plant widely used in traditional medicine for the treatment of diseases around the world, used for many years as a supplement and as medicine for its healing properties, reported in many religions and ancient mythologies and has recently been studied for its therapeutic properties in favor of human health care [13, 16, 17]. As currently available products containing the compounds present in pomegranate presented as tablets, capsules, tea or edible gels [5, 11, 17].

This fruit has various pharmacological properties demonstrated experimentally, thanks to the presence of phenolic compounds, such as antimicrobial, anti-inflammatory, antiviral, antiparasitic, antifungal, anti-hepatotoxics, anti-lipoperoxidant, antidiabetic, anticancer, and chemopreventative [9, 14, 16, 18], therefore it has been reported that pomegranate can be used for treatment or prevention of a large number of diseases, such as obesity, diabetes, stomach diseases, oral diseases, skin damage by UV radiation, infertility, Alzheimer's, arthritis, treatment against influenza virus, respiratory diseases, cardiovascular disease and various cancers such as prostate cancer, breast cancer, colon cancer, lung cancer and leukemia [1, 2, 5, 8, 11, 16, 19, 20].

**TABLE 5.1**    Compounds Present in Each Part of the Tree and Fruit [2, 9, 18, 26]

| Part of the tree | Compounds |
|---|---|
| Seed (oil) | Esterols and conjugated linoleic acids, punicic acid, campesterol, γ-tocoferol, 17- α-estradiol, coumestrol, sigmasterol, β-sitosterol y estriol |
| Peel, pericarp | Ellagitannins (punicalagin, punicalin, pedunculagin), flavonoids, flavones, gallic acid and ellagic acid. |
| Leaves | Tannins such as ellagitannins and flavone glycosides |
| Flowers | Ursolic acid, anthocyanins, tannins, and triterpenes |
| Root and bark | Ellagitannins and alkaloids |
| Arils | Vitamin c, b5, potassium, and phenolic compounds |

As a consequence of the properties and benefits of the compounds present in pomegranate peel, the use mainly in the food industry has been increasing significantly [1] however it is still underused, usually pomegranate is a fruit that is consumed fresh, although recently it has been used for the elaboration of many other products such as juices, alcoholic beverages and jams [21], although mainly juice, since this is the most popular and consumed product. Pomegranate juice is rich in phenolic compounds, with a high antioxidant capacity, higher in comparison to other products like wines and teas, this because the juice is elaborated with the extracts of the whole fruit, therefore in the last years there has been a lot of research focused on describe the components present on pomegranate juice [2, 7] who have been able to demonstrate the benefits of fruit and juice to the human health [11].

However, there is some doubt about the quality and if these remedies are safe and effective for the treatment of diseases [2], since these extracts are rich in several compounds that could be toxic to human health if they are consumed in large quantities or for long periods of time, which has triggered a huge interest in the scientific investigation of natural extracts, to demonstrate their safety before their use [2, 5]. It has been reported that these extracts have toxic properties, but due to the limited literature on this subject, this makes it difficult to understand the toxicity of pomegranate peel extracts [5, 16]. In order to validate these natural extracts, it is necessary to perform further tests to evaluate the results and identify the effects that these can cause, as well as the compounds present in the extract and perform a specific analysis of each molecule separately [2].

## 5.3　BIOTECHNOLOGY IMPORTANCE OF POMEGRANATE

There is evidence in the literature that there are about 124 different phytochemicals present in pomegranate, among which can be found ellagic acid and its derivatives, anthocyanidins, and hydrolyzable tannins such as ellagitannins and the most abundant ellagitannin in pomegranate, punicalagin. These compounds are mainly responsible for antioxidant and anti-inflammatory effects when consuming pomegranate. Hydrolyzable tannins, principally ellagitannins, are the compounds in pomegranate with the highest antioxidant activity among the other phenolic compounds present therein, among which are ellagic acid, punicalagin (responsible for more than 50% of the antioxidant activities), punicalin, and gallic acid. Although some studies *in vivo* suggest that the biological action depends on compounds which are metabolized from ellagitannins [1, 22]. Ellagic acid is a polyphenol present in a large variety of plants such as oak and eucalyptus, fruits (pomegranate, berries, grapes, etc.) and nuts [23] which is used as an additive in many foods and has also been documented as a compound widely used for the treatment of diseases such as cancer, for the reduction of accumulated fats and triglycerides in the body, has a cytoprotective effect on living cells with oxidative damage and damaged DNA [5, 24, 25].

Pomegranate contains a high variety of flavonoids distributed throughout the fruit, these compounds represent about 0.2–1.0% of the total weight of the fruit, of which 30% of the anthocyanidin's present in the fruit can be found in the shell. The concentration of these compounds is dependent and varies according to the type of crop of the plant and to the different phases of its development [1]. Pomegranate is one of a few plants known to contain conjugated fatty acids, this may relate to this fruit with its anticancer properties [17].

It is not only the edible fraction of the pomegranate that has biological properties, since the other parts of the fruit (root, leaves, seeds, stem, etc.) also have properties, among them antioxidants [10], and all parts can be used, some for the treatment of diseases, although most fruits have therapeutic properties, in recent years pomegranate has taken a lot of scientific relevance, thanks to its constituents and pharmacological mechanisms. Since all parts of the pomegranate contain important molecules (Table 5.1), this fruit is considered as a functional food [2]. Pomegranate is a rich source of minerals, vitamins, antioxidants like polyphenols and tannins. Compounds such as tannins are found in all parts of the tree, from the fruit to root [3]. The antioxidant activity of extracts of this fruit, both edible part, juice, and peel, is the reason for the enormous interest in this plant, just because to the amount of antioxidant compounds present in it, such as ellagitannins and flavonoids [10].

## 5.4 POMEGRANATE PEEL

The pomegranate peel represents approximately 50% of the total weight of the fruit, in which about 48 compounds have been identified and it is the peel which contains the largest quantity of phenolic compounds in relation to any other part of the fruit. In the peel it can be found a large variety of compounds, including flavonoids, hydrolyzable tannins, both gallotannins and ellagitannins (oligomers of punicalagin and punicalin, pedunculagin, ellagic acid, gallic acid), proanthocyanidins, anthocyanins (delphinidin, cyanidin, and pelargonidin), glucose esters, organic acids, proteins, moisture, and ashes. The ellagitannins, such as punicalin, punicalagin, and pedunculagin, together with glucosides of ellagic acid, are the compounds present in the largest quantity on pomegranate peel. The concentration of ellagic acid in peel and pomegranate juice is around 10–50 mg/100 g and 1–2.38 mg/100 mL, respectively [1, 5, 27]. The isomers of punicalagin are the major compounds present in the pomegranate peel (82.4 mg/g), but the concentration depends very much on the area where the plant is grown and process and storage conditions of the peel once dried [25].

As for the compounds in pomegranate peel extracts, they have nutraceutical relevance and together show antimutagenic, antioxidant, antimicrobial, antiinflamatory, and antibacterial effects against some pathogens (*Escherichia coli, Salmonella* spp. *Shigella* spp, *Vibrio cholerae*) [1, 2, 4, 5]. Extracts obtained from the pomegranate peel have been used to treat diseases such as diarrhea, dysentery, worms stomach invasion, ulcers, sore throat and dental problems. This part of the fruit has been used as a remedy for various diseases in different cultures over the years, as the Egyptian or Chinese culture, since a third of provinces in China grow this tree. These medical properties gradually awaken the interest of the scientific community, which has generated an increase in the search to understand the role of these compounds in human health. The bioactive potential of punicalagin is because this can be hydrolyzed to ellagic acid in the intestine [5, 25].

Nutritional, functional, and health-beneficial properties of pomegranate peel and its extracts are being used to be applied as food additives, functional food ingredients and bioactive components used in nutraceutical products. Therefore, the biomolecules present in the peel are trying to substitute synthetic agents present as additives or chemopreventives, but among the compounds that have been isolated from the pomegranate peel, only a few have been thoroughly investigated to determine their efficiency and use [1].

However, functional foods containing pomegranate peel extracts are not well accepted by the consumer as it does not contain a pleasant taste which reduces their sensory properties. On the other hand, there are more qualities used in food industries, such as the addition of color, conservation, and improvement in food quality to make these foods more attractive to consumers [4].

## 5.5 EXTRACTION OF COMPOUNDS FROM POMEGRANATE PEEL

Considered medicinal plants are an important source of compounds with therapeutic potential and have been used for a long time to date, for the extraction of compounds with great medical relevance, however since the extraction of compounds has some difficulties, in recent years industries have focused on the use of synthetic origin ingredients for the discovery of new active ingredients.

Medicinal plants produce a high number of secondary metabolites with great biological functions, which have not yet been fully studied. However, the trend indicates that these natural products will become important as sources of active components in the future. One of the limitations in the extraction of these compounds is the low yield that is obtained, which makes difficult its use as well as the methods of identification, whether genetic or chemical, since among many sources of bioactive molecules nowadays, it is necessary to have a classification of each ingredient for further analysis [15].

The recovery of phenolic compounds, specifically from the pomegranate peel at industrial levels, can be carried out using solvent-extraction processes, however, these procedures prove not to be environmentally friendly, among them water, methanol, ethanol, acetone, chloroform, and ethyl acetate, being the solvents with polar characteristics the ones that show a better performance in the extraction of antioxidant compounds.

The problem is the use of different solvents, because this can modify the properties or even the structure of phenolic compounds, since it has been reported that the use of methanol results in extracts with a higher anti-oxidant activity than extracts obtained with water, which show a greater antimutagenic activity [5, 28]

Some techniques for the extraction of phenolic compounds are maceration, agitation, extraction with aqueous solvents such as methanol and water, supercritical fluids, microwave-assisted extraction, but some of these methods have disadvantages due to high energy costs, pollution to the environment by the amount of waste generated by using high amounts of solvents [13, 28].

### 5.5.1  SOLID-STATE CULTURE

An alternative for the obtaining of different types of compounds is through solid-state culture (SSC); Pandey in 2003 defines SSC as a fermentation involving solid supports in absence or near absence of free water, but with enough moisture that allows microorganisms growth [29]. SSC has gained interest in recent years, mainly by the industrial sector that is focusing in the development and use of natural ingredients and additives, derived from microorganisms instead of synthetic products that carry a chemical process for its realization [30, 31].

The area of SSC has been growing, and with it the appearance of new bioprocesses in different work fields, such as bioremediation and biodegradation of hazardous compounds, biological detoxification of agro-industrial residues, biotransformation of crops for nutritional enrichment, biopulping, production of value-added products (Table 5.2.) [31], and the production of fermented foods (wine, soy sauce and vinegar) [30]. In addition, it is mentioned in the literature that a higher yield is obtained in production in SSC in comparison with submerged fermentation (SmF) [32].

The use of this technique has several advantages, among which is the low energy requirement, a higher product yields, less wastewater production, a lower risk of bacterial contamination, it is eco-friendly and one of the most important is the use of agroindustrial wastes as support and source of carbon for the fermentation [33]. The importance of using these agro-industrial wastes is to give an added value to these wastes that most of the time are not used and since they are not given an adequate treatment, when

**TABLE 5.2** Value-Added Products From Solid-State Culture

| Products generated from Solid-state Culture | Reference |
|---|---|
| Fuel, biopolymers, biosurfactants, biofertilizers, biopesticides | [33] |
| Biologically active secondary metabolites and enzymes | [34] |
| Ethanol, single-cell protein (SPC), organic acids, amino acids, additives, and supplements (antioxidants, flavors, pigments, and sweeteners) | [30] |
| Industrial chemicals and pharmaceutical products, including antibiotics | [31] |

using this type of material it helps to decrease the problem of pollution, not to mention that it is economically feasible since it often has no cost or has a very low cost, which is very attractive for bioprocessing [30, 31, 33], principally from the food industries. There are reports where some extracts or even parts of the fruit are used as a substrate for the fermentation process, such as cranberry pomace, creosote leaves extract, sugarcane bagasse, corn cobs, coconut husk, candelilla stalks and pomegranate peel [22, 35–37].

Pomegranate peel can function as a substrate for the biotechnological production of bioactive compounds because it is a rich hydrolyzable tannins as carbon source, which can be biotransformed into ellagic acid, involving the action of a recently discovered enzyme called ellagitannin acyl hydrolase or ellagitannase, which is one of the enzymes produced by SSC using pomegranate peel [35–37]. For commercial applications, the production of a wide variety of enzymes from microbial origin has been highly reported due to its many applications [32, 38]. Tannin acyl hydrolase, commonly named tannase, is an enzyme with potential biotechnological applications in the industrial zone, principally food and pharmaceutical industries. Increasing the importance of these residues as a fermentation material [38–40]. Different authors have pointed out other enzymes related to tannins biodegradation, such as xylanase and cellulase [41] and β-glucosidase [42]. SSC shows many biotechnological advantages, this at laboratory scale, the limitation is found at the moment of trying to do a scaling of this process [31].

## 5.6 CONCLUSIONS

In recent years, pomegranate peel has been a source of study due to its properties, mainly towards human health; being the source of natural bioactive compounds that can be extracted and used as food additives or ingredients from the use of agroindustrial wastes, the problem is the difficulties involved in these extraction processes, finding another way to obtain these phytochemicals has become very relevant, giving way too many studies focused on this topic. However, it has not been fully exploited, leaving a big field of opportunity for future research and waste utilization as a source of bioactive components.

## ACKNOWLEDGMENTS

Author René Díaz Herrera would like to thank the Food Research Department at Universidad Autónoma de Coahuila and CONACyT for the financial support.

## KEYWORDS

- biotechnology importance
- hydrolyzable tannins
- pomegranate peel
- pomegranate uses
- solid-state culture (SSC) technique
- submerged fermentation

## REFERENCES

1. Akhtar, S., Ismail, T., Fraternale, D., & Sestili, P., (2015). Pomegranate peel and peel extracts: Chemistry and food features. *Food Chem.*, *174*, 417–425. http://doi.org/10.1016/j.foodchem.2014.11.035.
2. Bhandari, P. R., (2012). Pomegranate (*Punica granatum* L). Ancient seeds for modern cure? Review of potential therapeutic applications. *Int. J. Nutr., Pharmacol., Neurol. Dis.*, *2*(3), 171–184. http://doi.org/10.4103/2231-0738.99469.
3. Chauhan, R. D., & Kanwar, K., (2012). Biotechnological advances in pomegranate (*Punica granatum* L.). *In Vitro Cell. Dev. Biol.*, *48*, 579–594. http://doi.org/10.1007/s11627-012-9467-7.
4. Fischer, U. A., Carle, R., & Kammerer, D. R., (2011). Identification and quantification of phenolic compounds from pomegranate (*Punica granatum* L.) peel, mesocarp, aril and differently produced juices by HPLC-DAD – ESI / MS n. *Food Chem.*, *127*(2), 807–821. http://doi.org/10.1016/j.foodchem.2010.12.156.

5. Ismail, T., Sestili, P., & Akhtar, S., (2012). Pomegranate peel and fruit extracts: A review of potential anti-inflammatory and anti-infective effects. *J. Ethnopharmacol.*, *143*(2), 397–405. http://doi.org/10.1016/j.jep.2012.07.004.

6. Melgarejo, P., Martínez-Valero, R., Guillamón, J. M., Miró, M., & Amorós, A., (1997). Phenological stages of the pomegranate tree (*Punica granatum* L.). *Ann Appl Biol.*, *130*, 135–140.

7. Mertens-talcott, S. U., Jilma-stohlawetz, P., Rios, J., Hingorani, L., & Derendorf, H., (2006). Absorption, metabolism, and antioxidant effects of pomegranate (*Punica granatum* L.) polyphenols after ingestion of a standardized extract in healthy human volunteers. *J. Agric. Food Chem.*, *54*, 8956–8961. http://doi.org/10.1021/jf061674h

8. Miguel, M. G., Neves, M. A., & Antunes, M. D., (2010). Pomegranate (*Punica granatum* L.): A medicinal plant with myriad biological properties – A short review. *J. Med. Plants Res.*, *4*(25), 2836–2847.

9. Kim, N. D., Mehta, R., Yu, W., Neeman, I., Livney, T., Poirier, D., & Lansky, E., (2002). Chemopreventive and adjuvant therapeutic potential of pomegranate (*Punica granatum*) for human breast cancer. *Breast Cancer Res. Treat.*, *71*, 203–217.

10. Kaur, G., Jabbar, Z., Athar, M., & Alam, M. S., (2006). *Punica granatum* (pomegranate) flower extract possesses potent antioxidant activity and abrogates Fe-NTA induced hepatotoxicity in mice. *Food Chem. Toxicol.*, *44*, 984–993. http://doi.org/10.1016/j.fct.2005.12.001.

11. Lan, X., Dongming, X., Fan, L., Wei, W., Lizhen, X., Lei, N., & Lijun, D., (2008). Effects of season, variety, and processing method on ellagic acid content in pomegranate leaves. *Tsinghua Sci. Technol.*, *13*(4), 460–465.

12. FAOSTAT-FAO (2010). Statistical database. Food and Agriculture Organization of the United Nations, Codex Alimentarius Commision: Tunis, Tunesia. <http:// www.fao.org>

13. Matharu, A. S., De Melo, E. M., & Houghton, J. A., (2016). Opportunity for high value-added chemicals from food supply chain wastes. *Bioresour. Technol.*, *215*, 123–130. http://doi.org/10.1016/j.biortech.2016.03.039.

14. Cavalcanti, R. N., Navarro-díaz, H. J., Santos, D. T., Rostagno, M. A., & Angela, M. A., (2012). Supercritical carbon dioxide extraction of polyphenols from pomegranate (*Punica granatum* L.) leaves: Chemical composition, economic evaluation and chemometric approach. *J. Food Res.*, *1*(3), 282–294. http://doi.org/10.5539/jfr.v1n3p282.

15. Atanasov, A. G., Waltenberger, B., Pferschy-wenzig, E., Linder, T., Wawrosch, C., Uhrin, P., & Stuppner, H., (2015). Discovery and resupply of pharmacologically active plant-derived natural products: A review. *Biotechnol. Adv.*, *33*(8), 1582–1614. http://doi.org/10.1016/j.biotechadv.2015.08.001.

16. Angel, S., Fonseca, G., Luis, J., Cozzi, R., Cundari, E., Fiore, M., & Salvia, R. De., (2008). Assessment of the genotoxic risk of *Punica granatum* L. (Punicaceae) whole fruit extracts. *J. Ethnopharmacol.*, *115*, 416–422. http://doi.org/10.1016/j.jep.2007.10.011.

17. Hora, J. J., Maydew, E. R., Lansky, E. P., & Dwivedi, C., (2003). Chemopreventive effects of pomegranate seed oil on skin tumor development in CD 1 Mice. *J. Med. Food.*, *6*(3), 157–161.

18. Mehta, R., & Lansky, E. P., (2004). Breast cancer chemopreventive properties of pome-granate (*Punica granatum* L.) fruit extracts in a mouse mammary organ culture. *Eur. J. Cancer Prev.*, *13*, 345–348. http://doi.org/10.1097/01.cej.0000136571.70998.5a.

19. Chinsembu, K. C., (2016). Plants and other natural products used in the management of oral infections and improvement of oral health. *Acta. Trop.*, 154, 6–18.

20. González-Ortiz, M., & Martínez-Abundis, E., (2011). Effect of pomegranate juice on insulin secretion and sensitivity in patients with. *Nutr. Metab.*, *58*, 220–223. http://doi.org/10.1159/000330116.

21. Hernández, F., Melgarejo, P., Olías, J. M., & Artés, F., (2000). Fatty acid composition and total lipid content of seed oil from three commercial pomegranate cultivars, *Adv. Res. Technol.*, *209*, 205–209.

22. Patel, C., Dadhaniya, P., Hingorani, L., & Soni, M., (2008). Safety assessment of pomegranate fruit extract: Acute and subchronic toxicity studies. *Food Chem Toxicol.*, *46*(8), 2728–2735. http://doi.org/10.1016/j.fct.2008.04.035.

23. Aguilera-Carbo, A., Hernández, J. S., Augur, C., Prado-Barragan, L. A., Favela-Tor-res, E., & Aguilar, C. N., (2009). Ellagic acid production from biodegradation of *Creosote bush* ellagitannins by *Aspergillus niger* in solid-state culture. *Food Bioprocess Technol.*, *2*(2), 208–212. http://doi.org/10.1007/s11947-008-0063-0.

24. Ascacio-Valdés, J., Aguilera-Carbó, A., Buenrostro, J., Prado-Barragán, A., Rodrí-guez-Herrera, R., & Aguilar, C. N., (2016). The complete biodegradation pathway of ellagitannins by *Aspergillus niger* in solid-state fermentation. *J. Basic Microbiol.*, *56*, 329–336. http://doi.org/10.1002/jobm.201500557.

25. Lu, J., Ding, K., & Yuan, Q., (2008). Determination of punicalagin isomers in pome-granate husk. *Chromatographia.*, *68*(3), 303–306. http://doi.org/10.1365/s10337-008-0699-y.

26. Kohno, H., Suzuki, R., Yasui, Y., Hosokawa, M., Miyashita, K., and Tanaka, T. (2004). "Pomegranate seed oil rich in conjugated linolenic acid suppresses chemi-cally induced colon carcinogenesis in rats," *Cancer Science*, 95 (6), pp. 481–486.

27. Aviram, M., Volkova, N., Coleman, R., Dreher, M., Kesava, M., Ferreira, D., & Rosenblat, M., (2008). Pomegranate phenolics from the peels, arils, and flowers are antiatherogenic: Studies *in vivo* in atherosclerotic apolipoprotein E-deficient (E0) mice and *in vitro* in cultured macrophages and lipoproteins. *J. Agric. Food Chem.*, *56*, 1148–1157.

28. Wong-paz, J. E., Muñiz-márquez, D. B., Aguilar-zarate, P., Rodríguez-herrera, R., & Aguilar, C. N., (2014). Microplate quantification of total phenolic content from plant extracts obtained by conventional and ultrasound methods. *Phytochem. Anal.*, *25*(5), 439–44. http://doi.org/10.1002/pca.2512

29. Pandey, A., (2003). Solid-state fermentation. *Biochem. Eng. J.*, *13*(2–3), 81–84. http://doi.org/10.1016/S1369-703X(02)00121-3.

30. Couto, S. R., & Sanromán, M. Á., (2006). Application of solid-state fermentation to food industry-A review. *J. Food Eng.*, *76*(3), 291–302. http://doi.org/10.1016/j.jfoodeng.2005.05.022

31. Singhania, R. R., Patel, A. K., Soccol, C. R., & Pandey, A., (2009). Recent advances in solid-state fermentation. *Biochem. Eng. J.*, *44*(1), 13–18. http://doi.org/10.1016/j.bej.2008.10.019.

32. Aguilar, C. N., Gutiérrez-Sánchez, G., Prado-Barragán, L. A., Rodríguez-Herrera, R., Martínez-Hernández, J. L., & Contreras-Esquivel, J. C., (2008). Perspectives of solid-state fermentation for production of food enzymes. *Am. J. Biochem. Biotechnol.*, *1*(4), 354–366. http://doi.org/10.1017/CBO9781107415324.004.

33. Thomas, L., Larroche, C., & Pandey, A., (2013). Current developments in solid-state fermentation. *Biochem. Eng. J.*, *81*, 146–161. http://doi.org/10.1016/j.bej.2013.10.013

34. Hölker, U., & Lenz, J. (2005). Solid-state fermentation—are there any biotechnological advantages?. *Current Opinion in Microbiology*, *8*(3), 301-306.

35. Ascacio-Valdés, J. A., Buenrostro, J. J., De la Cruz, R., Sepúlveda, L., Aguilera, A. F., Prado, A., & Aguilar, C. N., (2013). Fungal biodegradation of pomegranate ellagitannins. *J. Basic Microbiol.*, 54(1), 28–34. http://doi.org/10.1002/jobm.201200278.

36. Buenrostro-Figueroa, J., Huerta-Ochoa, S., Prado-Barragán, A., Ascacio-Valdés, J., Sepúlveda, L., Rodríguez, R., et al., (2014). Continuous production of ellagic acid in a packed-bed reactor. *Proc Biochem.*, *49*(10), 1595–1600. http://doi.org/10.1016/j.procbio.2014.06.005.

37. Robledo, A., Aguilera-Carbó, A., Rodriguez, R., Martinez, J. L., Garza, Y., & Aguilar, C. N., (2008). Ellagic acid production by *Aspergillus niger* in solid-state fermentation of pomegranate residues. *J. Ind. Microbiol. Biotechnol.*, *35*(6), 507–513. http://doi.org/10.1007/s10295-008-0309-x.

38. Gayen, S., & Ghosh, U., (2013). Purification and characterization of tannin acyl hydrolase produced by mixed solid-state fermentation of wheat bran and marigold flower by *Penicillium notatum* NCIM 923. *BioMed Res. Int.*, 1–6. http://doi.org/10.1155/2013/596380.

39. Fuentes-Garibay, J. A., Aguilar, C. N., Rodríguez-Herrera, R., Guerrero-Olazarán, M., & Viader-Salvadó, J. M., (2015). Tannase sequence from a xerophilic *Aspergillus niger* strain and production of the enzyme in *Pichia pastoris*. *Mol. Biotechnol.*, *57*(5), 439–447. http://doi.org/10.1007/s12033-014-9836-z.

40. Ramos, E. L., Mata-Gómez, M. A., Rodríguez-Durán, L. V., Belmares, R. E., Rodríguez-Herrera, R., & Aguilar, C. N., (2011). Catalytic and thermodynamic properties of a tannase produced by *Aspergillus niger* GH1 grown on polyurethane foam. *Appl. Biochem. Biotechnol.*, *165*(5–6), 1141–1151. http://doi.org/10.1007/s12010-011-9331-y.

41. Huang, W., Niu, H., Li, Z., He, Y., Gong, W., & Gong, G., (2008). Optimization of ellagic acid production from ellagitannins by co-culture and correlation between its yield and activities of relevant enzymes. *Bioresour Technol.*, *99*(4), 769–775. http://doi.org/10.1016/j.biortech.2007.01.032.

42. Vattem, D. A., & Shetty, K., (2003). Ellagic acid production and phenolic antioxidant activity in cranberry pomace (*Vaccinium macrocarpon*) mediated by *Lentinus edodes* using a solid-state system. *Proc. Biochem.*, *39*(3), 367–379. http://doi.org/10.1016/S0032-9592(03)00089-X.

# CHAPTER 6

# TECHNOLOGICAL ADVANCES IN THE PRODUCTION OF PHYTASES

JOSÉ DANIEL GARCÍA,[1] GEORGINA MICHELENA ÁLVAREZ,[2]
ERIKA NAVA-REYNA,[3]
JANETH MARGARITA VENTURA SOBREVILLA,[1] ANNA ILYINA,[1]
and JOSÉ LUIS MARTÍNEZ HERNÁNDEZ[1]

[1] *Food Research Department, Autonomous University of Coahuila, Mexico, E-mail: jose-martinez@uadec.edu.mx*

[2] *Cuban Institute for Research on Sugarcane Derivatives, Cuba*

[3] *National Institute for Forestry, Agriculture, and Livestock Research (INIFAP) CENID-RASPA, Mexico*

## ABSTRACT

Due to high consumption of phosphorus in agriculture and animal feeding, the improvement of techniques to get phosphorus has taken priority. Moreover, grains are the most common components in animal feeding, where phosphorus is usually found as phytates that are unavailable nutrients for monogastric animals. Therefore, phytases research has increased in the latest years to find better production conditions that might affect the cost, reduce production steps or increase yields.

Phytases hydrolyze phytate and release the phosphorus allowing the use of this macronutrient. Thermostable phytases have been developed because of the need for stable enzymes in processes that involve high temperatures and different pH conditions similar to the gastric or intestinal conditions where it will be used. Nowadays, the optimization of phytases production is one of the main issues that are looked for through the improving of the culture, the bioreactor, microorganisms, etc. It will enhance the storage availability and will obtain better performance at high temperatures and different pH's. A stable enzyme can be achieved using additives as maltose, glycerol, casein, polyethylene glycol, etc. In fact, to obtain these thermostable enzymes, genetics has an important role in using vectors to insert genes in microorganisms to get more resistant enzymes. The aim of this chapter is to review the recent researches, the technological advances and the improved processes in the actual production of phytases.

## 6.1 INTRODUCTION

Animal feeding is very important for livestock industry; animal diets need to provide all the requirements (like vitamins and minerals) for the optimal development of animals. Therefore, research is focused on developing food and food additives that fit those nutrient requirements. One of the main nutrients is phosphorus; after proteins and energy, it is the third most important component in animal diets and it represents the highest cost in animal feeding [1]. Cereals, the main component in animal diets, contain phosphorus as phytates, which cannot be digested by monogastric animals. Moreover, phytates are considered as antinutrient factors and diets must be complemented with external inorganic phosphorus, which represents an extra cost, as well as the undigested phytates in animal feces could be hydrolyzed by soil microorganisms and, subsequently, the free phosphorus may be lixiviated to aquifers and conducted to freshwater bodies where it is involved in eutrophication [2]. One strategy to avoid this situation is the use of natural phosphorus in cereals adding enzymes to hydrolyze phytates. These enzymes are named phytases. Since 1990 some researches have been developed to enhance the use and efficiency of phytases in animal's diet, hydrolyzing phytates to produce phosphorus and myo-inositol, and consequently, increase the bioavailability of minerals as phosphorus. According to Rodehutscord, phosphorus' availability is the total part of dietary phosphorus that, at minimal quantity of supply, can be used to cover animals' requirement of this mineral [3]. High phosphorus demand around the world, has generated a widely industrial growing. Moreover, the BM journal editor estimates a price of $250 million dollars in this field with an increase of 10–15% per year. This shows the great opportunity for research in this field. There are few industries producing phytases, therefore, research in this area has taken a very important place.

## 6.2 PHOSPHORUS

Phosphorus is one of the most essential nutrients in the development of bones and their metabolic processes [4]. Close to 0.83–9.15% is found in cereals like corn, wheat, sorghum, barley, etc., where it is found mainly as phytates [5]. Furthermore, phytates are considered as a polluting agent in farm areas. Several techniques to extract phytates from cereals and legumes have been reported, such as add sodium hydroxide, formaldehyde, ammonia, tannic acid and lactic acid. It has been shown that the use of lactic acid promotes the hydrolysis of phosphorus and in this way improves the absorption of this same element, which shows that the use of organic acids offers a good alternative to guarantee the necessary requirements and avoid environmental contamination [6]. However, most of these processes tend to be laborious and in some cases, they bring hazards to health.

Another problem is the size of food particles, since very fine food particles reduce the digestion of nutrients, including phosphorus [7]. Likewise, the importance of the size of the particles has been mentioned for the availability of its nutrients. When food is found to be used like pellets or recently ground and administered to animals, it is recommended a size of 600 and 900 μm to get greater benefits [8]. Thus, there is a great research opportunity in this area looking forward to improving the absorption of these nutrients.

The importance of these considerations is that phosphorus is involved in bone formation, energy reserve, cell signaling and stabilization of cell membranes. A deficit in phosphorus requirement may develop some diseases in animals. For example, bone disruption occurs because of the deficiency of nutrients such as calcium, vitamin D and phosphorus, which is reflected in the economic losses and are indeed harmful to the animal [9].

This situation has led to the addition of inorganic phosphorus from rocks that are mainly extracted by China, the United States and Morocco. Lwin et al. [2] conducted a study with the purpose of estimating the possible changes of phosphorus worldwide by simulating different scenarios. It was observed that by 2100, India, China, Brazil, and the United States would present the major phosphorus consumption per year. Cordell et al. and De Haes & Smith in 2009 [10, 11] also estimated that within 50 to 100 years we will be running out of these resources because of its wide use in agriculture and other processes, which will cause a great lack of phosphorus by then. As we see this, we can realize the need to develop processes that allow the utilization of natural phosphorus from cereals, such as phytates, for monogastric animals feed [12]. The uses of phytases (enzymes that hydrolyze phytate) are an alternative to get this aim.

## 6.3   PHYTIC ACID AND PHYTATE

Only 33% of total phosphorus content in cereals, wheat, triticale, rice, sorghum, soybeans, corn, etc., [13] is used because they have anti-nutritional factors such as phytic acid, an organic acid composed of 6 phosphate molecules $(PO_4^-)$ bound with a molecule of myoinositol with a pKa of 4.6 to 10, which fulfills functions in plants like energy reserve, source of phosphorus, antioxidant, and precursor of the cell membrane among others [13]. Owing to its pKa, it has a chelating effect, allowing it to bind minerals as Ca+, Na+ and Mg+, as well as proteins and sugars, the reason why these compounds can inhibit enzymes [14]. These unions form complexes known as phytates, which cannot be assimilated by monogastric animals [15]. It should be emphasized that studies have shown that phytates have also anticancer effects and have been used as agents in the prevention of breast, colon, prostate cancer; even more, phytates have been applied against HIV, diabetes, dental caries, to mention a few [16].

Phosphorus used to solve this problem is mainly obtained from rocks [16, 17] that involves a high cost to obtain this mineral. Therefore, the search for low-cost alternatives that allow the natural use of this mineral present in cereals has increased.

## 6.4   PHYTASES

Phytases are a group of enzymes belonging to acid phosphatases of the hydrolases type, which hydrolyze the monoester bond releasing phosphate and myoinositol groups. The first reports of these enzymes date back to 1907 by Suzuki et al. [19], but it was not until 1990 that the first enzyme was isolated and characterized using *Aspergillus ficuum* [1]. Phytases are composed approximately of 27.3% oligosaccharides, 37% non-polar amino acids, 42% polar amino acids, 9.5% acidic amino acids and 9.5%

alkaline amino acids. Within its secondary structure are 17.3% α-helices, 29.1% β-folds, among others [20].

## 6.4.1 CLASSIFICATION OF PHYTASES

Phytases can be classified according to the source from which they come, so we have phytases of microbial, vegetal, and animal origin [21] (see Table 6.1). Bacterial phytases have proved to be a better alternative to the fungal phytases because of some properties such as their catalytic efficiency, substrate specificity and resistance to proteolysis [12, 22], however, bacterial phytases are not widely used at industry due to their lower resistance [23].

Phytases can also be classified according to their optimal pH, having in this way alkaline and acidic phytases. For example: phytases of *Aspergillus niger* have their optimal pH range between 2.5–5.5 while *B. amyloliquefaciens* phytase has a pH range of 7.0–8.0 [15].

Another classification is based on the site in which phytic acid binds to myo-inositol. The International Union of Pure Applied Chemistry and International Union of Biochemistry have called these enzymes as 3-phytase, 5-phytase, and 6-phytase [24].

**TABLE 6.1** Phytases Classification

| According to | Type | Characteristic |
|---|---|---|
| Source | Microbial | Bacteria, fungi, and yeasts |
| | Vegetal | Cereals: corn, wheat, sorghum, etc. |
| | Animal | Ruminants |
| pH | Alkaline | pH around 8 |
| | Acids | pH 3.0–5.5 |
| NC-IUBMB* | 3-phytase | Per the number of carbon that is attached to the myo-inositol |
| | 5-phytase | |
| | 6-phytase | |
| Others | Exogenous | Mostly from microbial (*Aspergillus* and *Saccharomyces*) |
| | Endogenous | From some cereals (wheat and legumes) and from intestinal flora (large intestine of swine) |

* The International Union of Pure Applied Chemistry and the International Union of Biochemistry. Source: [18, 24, 25].

## 6.4.2 MECHANISM OF REACTION

Phytases act as a monomer with two domains: an alpha domain and an alpha/beta domain at the active site of these enzymes, where the catalytic reactions are carried out, even though both domains are involved. Determination of crystalline structure through X-rays is a technique that allows us to determine the spatial arrangements of atoms that are part of a molecule [25]. This technique has been used to know the phytases structures. For example, Oakley [26] in 2010 characterized the phytase of *A. niger,* finding that it is formed by an α-domain, an α/β domain and an N-terminal. Ha [27] in 1999 also reported one of the first structures of these enzymes using this methodology. The mechanism of the enzyme coming from *Bacillus subtilis* is related to what was mentioned before by the way in which these enzymes act through the domains presented in the enzyme by two phases in which the presence of the covalent union of phospho-histidine is required. This intermediate will be used for the transfer of the phosphoryl group by a nucleophilic attack of a phosphate group which undergoes cleavage by a histidine residue and the protonation of a leaving group by another one present on the enzyme. This results in a covalent phospho-enzyme complex and an alcohol molecule. The complex formed is very unstable, leading to hydrolysis and producing inorganic phosphorus and a molecule of myoinositol [28] (see Figure 6.1).

$$E + ROPO_3H^- \longleftrightarrow E \cdot ROPO_3H^- \xrightarrow{ROH} E \cdot PO_3^- \xrightarrow{H_2O} E + Pi$$

**FIGURE 6.1** Scheme of the hydrolysis reaction of phytase on phytate.

It has been reported that the mechanism by which phytases release phosphorus is characteristic for each enzyme, *A. niger* shows two different types of phytase enzymes, A and B, and for each one the mechanism is unique [20].

## 6.4.3 THERMOTOLERANT PHYTASES

Since the research on phytases began, the interest of the researchers has been focused on the production, getting active enzymes generally at low pH and resistant to low temperatures, which represents a problem since enzymes are used in food where temperatures above 40°C are applied to grind cereals, besides the instability at neutral pH. Due to this, enzymes production must be increased and it was consequently reflected in its cost and yield. This led to the search for thermotolerant phytases (see Table 6.2). Singh & Satyanarayana in 2009 [29] evaluated the stability of the phytase produced by *Sporotrichum thermophile* to high incubation temperature (from 60 to 80°C), and variation in pH (3.0, 5.0 and 7.0), finding that the phytase produced by this organism has a weight of 90 kDa, retains an activity of 100% at a temperature of 60°C and is active to acid pH's.

**TABLE 6.2**    Thermotolerant Phytases

| Microorganism | Molecular Weight | pH | Temperature, °C | Activity | Reference |
|---|---|---|---|---|---|
| *Aspergillus ficuum* | 65 kDa | 1.3 | 67 | 150 U/mg | [29] |
| *Rhizopus oryzae* | 34 kDa | 1.5–2.0 | 45 | 141.83 U/mg | [30] |
| *Schizophyllum commune* | NA | 5 | 50 | 133.76 U/gds | [31] |
| *A. oryzae* | NA | 5 | 30 | 145 U/g | [32] |
| *H. nigrescens* | NA | 5 | 45 | 44.86 U/gds | [33] |
| *Sporotrichum thermophile* | 55 kDa | 5 | 60 | NA | [34] |
| *E. coli* | 42 kDa | 2 | 55 | 363 U/mg | [35] |

NA: Not available; gds: grams dry substrate.

Gaind and Singhin 2015 [37] isolated an *A. flavus* phytase finding characteristics like 80% phytase activity and optimum pH of 6.0–37° C with a molecular weight of 30 kDa. Sato et al. [38] characterized the thermotolerant phytase produced by *Rhizopus microspurs* by fermentation with polyethylene biofilms, obtaining a phytase with molecular weight of 35.4 kDa with a favorable activity at pH 4.5. However, this same enzyme has phytase activity under alkaline conditions at pH 8.0–10.5 with an activity of 65–75%, which represents a great contribution because there are few records of phytases with activity at this pH range. Maurya et al. [39] used *Escherichia coli* and *Picha Pastori* as vectors to clone the rStPhy gene isolated from *S. thermophile* and thereby achieved a thermotolerant enzyme by using two promoters: alcohol oxidase (AOX) and phosphate glyceraldehyde dehydrogenase (GAP). They observed that the AOX promoter was affected by the presence of glucose and glycerol, presenting risks with methanol in industrial concentrations and taking too much time in fermentation that is better found with the GAP promoter, since it does not need the use of methanol for its activation, which made it more competent. AOX produced 480 U/ml$^{-1}$ while GAP produced 495 U/ml$^{-1}$. Therefore, with the observed advantages, GAP results to be the most effective promoter. The result was an enzyme obtained with 95% of phytase activity at 60°C and pH 5.0.

Yu and Chen [40] report a good alternative of phytase enzymes, because most of the reported thermotolerant phytases are derived from fungi: a thermotolerant phytase from the bacterium *Bacillus nealsoni* that exhibits 73% activity at 55°C in a pH range from 6.0 to 8.0, where 7.5 shows the highest production [41].

## 6.5 PHYTASE PRODUCTION

### 6.5.1 PHYTASE PRODUCTION METHODS

In ancients times fungi were cultivated in a solid-state fermentation. An example of this is the koji process by *A. oryzae* [42], where long fermentation period of time is required to obtain different kind of products. Since the beginning of these researches in 1990, the first phytases isolated were carried out by a solid-state fermentation; solid, semi-solid or liquid fermentation type have been evaluated over the years [18, 40–43]. Depending on the necessary requirements and considering the advantages and disadvantages of each type of fermentation, the correct type of fermentation will be chosen for the production of phytases. Some advantages offer by solid-state fermentation are lower plant operational cost, improved product recovery, simple cultivation equipment and high product concentration [47]. Several studies have compared the same microorganism in two types of fermentation: solid and liquid. It was observed that solid-state fermentation has a higher production of phytases [48] which can be attributed to the fact that this process simulates the natural state of fungus growth since it provides better-growing conditions. Comparing the results of this production, it is stated that between liquid state or submerged state and the solid-state fermentation, the solid-state production comes to be almost ten times more productive than liquid, which can be seen in the optimization probes that are almost made in solid matrix. In 2016, Kanti [49] compared two organisms and two types of fermentation in order to observe which one had the best production. *Neuroespora crassa* and *Neuroespora sithophila* were the microorganisms used in submerged fermentation and solid-state fermentation. Once again, solid-state fermentation showed better results with *N. crassa* and these results had not been reported before with this microorganism, which gives us new options for phytase production. Therefore, solid-state fermentation is proposed as the most viable for phytases production, due to high yields and low-cost, because the supports used for these products are mostly low-cost agroindustrial residues [41].

*A. niger and A. ficuum* were assessed in two kinds of fermentation by Shivanna [50]. Solid-state fermentation using wheat bran was better than submerged one, besides of a mix of wheat bran, rice bran, and groundnut cake in the ratio of 2:1:1, the outcome was enhanced in a 20 per cent more than the first one and, in contrast with the submerged fermentation, this mix showed an improvement of eight-fold higher than those results found with a submerged fermentation grown in potato dextrose broth. It should be noted that the ideal mix can improve the enzyme level production at low-cost (see Table 6.3).

### 6.5.2   IMPROVEMENTS IN PHYTASE PRODUCTION

#### 6.5.2.1   CONDITION AND CULTURE MEDIA

The optimization process has promising objectives and can focus on various outcomes such as its cost, its productivity, the improvement of its components as well as products with additional properties as the thermotolerant phytases mentioned above (see Table 6.4).

One of the optimization components is the substrate, which is obtained using industrial waste that, at some point, represent a pollution problem when it is not used and is also later discarded. Kumari et al. [41] in 2016 applied the microorganism *S. thermophile* to produce phytases using sugarcane bagasse as a substrate with 3% glucose as carbon source and 0.5% ammonium sulfate as nitrogen source, obtaining a reduction in time production from 120 h to 45 h with a productivity of 2906.3 $Ukg^{-1}h^{-1}$ to 39,187.50 $Ukg^{-1}h^{-1}$ at a temperature of 45°C.

**TABLE 6.3** Phytases Produced by Microorganisms and Some of Their Recovery Conditions

| Microorganism | Substrate | Carbon source | Nitrogen source | Enzymatic activity | Fermentation | Reference |
|---|---|---|---|---|---|---|
| *Neuroespora crassa* | Maize, beans, rice flour | Starch | Peptone | 45.25 U/gs | SSF | [46] |
| *Neuroespora sithophila* | Maize, beans, rice flour | Starch | Yeast extract | 40.78 U/gs | SSF | [46] |
| *Aspergillus niger* | Rice bran | Glucose | Ammonium nitrate | 17.8 U/ml | SmF | [45] |
| | | | | 28.6 U/ml | SSF | |
| *Pichia anomala* | Wheat bran | Glycerol | Yeast extract | 756 U/ml | | [81] |
| *Aspergillus niger* | Cane molasses | NA | Ammonium nitrate | 0.617 U/ml | SmF | [55] |
| *Aspergillus oryzae* | Wheat bran | Sucrose | Meat extract | 145 U/gs | SSF | [32] |
| *Ganoderma sp MR–56* | Wheat Bran Broth | NA | Yeast extract | 14.5U/ml | SmF | [60] |
| *Sporotrichum thermphile* | Sesame | Glucose | Ammonium sulfate | 349 U/gs | SmF | [40] |
| *Lactobacillus casei* | Whole wheat flour | NA | NA | NA | NA | [82] |
| *Aspergillus flavus* | Indian Mustard | Glucose | Malt extract | 112.25 U/gs | SSF | [36] |
| *Rhizopus oryzae* | Linseed | Mannitol | Ammonium Sulphate | 141.83 U/mg | SSF | [30] |

NA: Not available; SSF: Solid-state fermentation; SmF: Submerged fermentation; gs: grams substrate.

*Aspergillus niger* is one of the most commonly used fungi, which can be seen in the reported studies in order to improve existing methodologies [47–53]. Sridevi in 2015 [62] optimized crop conditions using industrial waste at different temperatures finding that rice bran is an excellent substrate and also adding its economic advantages and increased production by enriching the medium with glucose and ammonium nitrate, as carbon and nitrogen sources, at pH 5.0 and 60°C by solid-state fermentation. Likewise et al. in 2015 [52], optimized the medium using *A. niger* in cane molasses as a substrate in different concentrations, 6–10% of them resulting in a higher concentration of 8% v/v of substrate in the enzyme recovering by means of submerged fermentation. In addition to this, the solid-state fermentation with cane sugar bagasse, 5, 7.5, 10, and 15 ratio of 1:2.5 was evaluated, being the concentration of 7.5 the most productive. The cane molasses reported as a substrate by N. Singh presents some advantages such as its low-cost that work as a source of nitrogen containing also vitamins, which makes it a better option for this process.

Another fungus used is *Aspergillus oryzae*, which was reported by Sapna & Singh in 2014 [33] to optimize also by solid-state fermentation with wheat bran as substrate, supplemented with sucrose, starch, $FeSO_4$, $MgSO_4$, and KCl at pH 5. They also found that adding a surfactant increased the productivity of the process. The surfactant evaluated was Triton X–100, which increases the permeability of the membrane and the obtention of more phytases.

Another substrate used to produce pytases is canola meal [43, 56]. Nava tested triticale as a substrate to obtain phytase, which has not been reported for this use. Despite there is some data of endogenous phytase activity, it has not been used as a substrate in phytase production [64]. Nava confirmed the use of triticale as a substrate due to the fact that they obtained an active enzyme in pH 5.5 and temperature of 55°C [65]. Neira improves this process by analyzing two triticale phenological stages and two mixtures of these phenological stages evaluating its effectiveness in the production of phytase. The results obtained were better with cereal crop phenology stage. The phytase obtained showed phytase activity of about 6 U/gds. This process was supplemented with glucose, yeast extract and potassium chloride [66].

**TABLE 6.4**	Optimization of Processes for Phytases Production

| Microorganism | Before | After | System | Substrate | Reference |
|---|---|---|---|---|---|
| *Schizophyllum commune* | 40.7 U/g ds | 113.7 U/gds | SSF | Wheat bran | [31] |
| *Aspergillus oryzae* | 5 U/gds | 145 U/gds | SSF | Wheat bran | [32] |
| *Penicillium purpurogenum* | 401 U/gds* | 444 U/gds | SSF | Corn | [61] |
| *Escherichia coli* | 226.74 U/mg | 498.47 U/mg | SmF | NA | [54] |
| *Aspergillus flavus* | 34.0 U/g | 112.25 U/g | SSF | Indian Mustard | [36] |

NA: Not available; SSF: Solid-state fermentation; SmF: Submerged Fermentation; gds: grams dry substrate. *: The data corresponds to the statistical result of the expected activity.

## 6.5.2.2	BIOPROCESS OF PHYTASE PRODUCTION

Bioreactors are very important for fermentation process, because it is here where the process is carried out. Salmon et al. [53] reported the use of *Ganoderma sp* MR–56 using wheat bran as a substrate, obtaining an enzyme effective at 30°C and pH 3.0 by submerged fermentation with wheat bran broth and yeast extract, at pH 6.0 and 30°C. The above parameters were established for the evaluation of the type of bioreactor necessary for the fermentation, using three types of bioreactors; Erlenmeyer flasks, Dreschel bottles and a stirred tank reactor, for which they observed that the stirred tank reactor presents better results, besides that it is the first report of the use of this kind of bioreactors by submerged fermentation. This gives us a perspective of future research that can be performed with the purpose of optimizing phytase production.

## 6.5.3	IMPROVEMENT WITH GENETIC ENGINEERING

Genetic engineering allows us to obtain conditions to improve the production of a process from the genetic point of view modifying organisms to improve the conditions of production. Due to the fact that plants produce phytases, there have been made studies about their genetical modification to increase vegetal phytase production; however, these modifications have not overtaken others sources of phytases [48, 49].

Microorganisms have also been genetically modified through the selection of genes coding for the production of phytases, their subsequent amplification and expression. Hao Tan et al. [70], expressed the gene rAppA_Gw in *E. coli* in 2015 whereby obtained an enzyme with an optimum pH 2.0 and temperature of 45°C, which is a valuable contribution due to its pH value, quite optimum. Hao Tan et al. [70], also worked with *Aspergillus ficuum* that produces an active enzyme at pH 1.3 [30]; nevertheless, compared to the enzyme produced by modified *E. coli*, *A. ficuum* showed lower activity.

Erpel et al. [71] used the mE228k gene of *Aspergillis niger* which was converted into a plasmid and introduced by biobalistics into cells UTEX–90 and cwl5 for the use of the algae *Chlamydomonas reinhardtii*. Expression of this gene was reflected by codon amplification corresponding to phytases in 51.6%, thus producing a high amount of enzyme and also taking advantage of the nutritional properties of the alga as it contains lipids, carbohydrates, and proteins [72], avoiding the addition of supplements. Chi-Wei Lan et al. [62] cloned the pET23b gene and expressed it in *E.coli* to evaluated parameters that interfere with phytase production. They obtained a considerable phytase production from 226.74 U/mg$^{-1}$ to 498.47 U/mg$^{-1}$. Therefore, these results show also a promising area for research that will generate new scientific contributions.

## 6.6   PURIFICATION AND CHARACTERIZATION OF PHYTASES

After producing phytases, a new stage is presented: the purification and characterization of the enzymes produced in order to have a phytase solution with a higher enzyme concentration whose steps are as important as its production. It is also important to get to know and identify the type of obtained phytase. Between purification processes by salts, salting out is one of the strategies mainly applied to purify proteins by means of ammonium sulfate [40], usually accompanied by another purification technique such as ultrafiltration or chromatography; however, this first purification technique achieves the elimination of some minerals and other substances of low relevance.

Ultrafiltration is a filtration process in which some pressure is applied to a liquid, generally nitrogen to avoid oxidation in the product, on a semipermeable membrane where the water and solutes of low molecular weight will be filtered, remaining in the retained liquid proteins of higher molecular weight than the size of the membrane used [52, 63]. Some of the considerations for the purification of proteins might be the size of the protein to choose the size of the membrane and, in this way, identify the portion of product containing the protein of interest. Phytases have been obtained by this type of filtration, achieving purifications of 34.72 fold [37]. Evaluating the process of concentration and purification of the phytases, a considerable change in concentration by means of this technique can be observed during the first 20 minutes of purification and a consequent decreasing in concentration after those first minutes mentioned, which may mainly be due related to viscosity of the liquid remainder [73], and also because it can plug up the membranes used. This factor must be considered to perform this type of filtration. In this way, the selection of appropriate conditions is important to prevent the loss of enzyme activity, because it is believed that ultrafiltration may cause this loss, due to the shear force applied in the system and the removal of ions that may have a positive effect on the enzyme activity [53, 54].

Chromatography is another process that can be used for the purification or separation of the proteins, since it allows the separation of the samples. In some cases, it is used after ultrafiltration and can combine several types of chromatography. Different types of chromatography for phytase purification have been evaluated, where it is seen that gel filtration chromatography using a Superdex 75 column has a higher performance compared to ion exchange chromatography [75].

Once phytase has been purified, the obtained enzyme can be characterized by evaluating the molecular weight through a zymogram. Some phytases of a large range with a molecular weight of 34 to 200 kDa [76, 77] have been reported (see Table 6.2). In some cases, the high molecular weight may be due to the fact that phytases are made up of dimers [30]. Temperature and pH are optimal data to the study of the working conditions of the enzyme, and both depend on the type of applied microorganism. Some examples of obtained enzymes and their optimal pH and temperature are in Table 6.5. It has also been reported that phytases have an isoelectric point of 3.65 and 5.2 for *Aspergillus niger* phytase.

**TABLE 6.5**   Temperature and pH Optimal Conditions for Phytases

| Microorganism | Temperature (°C) | pH | Reference |
|---|---|---|---|
| R. oryzae | 45 | 1.5 & 2.5 | [30] |
| Bacillus nealsonii | 55 | 7.5 | [39] |
| Burkholderia sp. | 44–55 | 4.5 | [71] |
| Aspergillus flavus | 45 | 7.0 | [36] |
| Aspergillus ficuum | 67 | 1.3 | [29] |
| Klebsiella pneumoniae | 50 | 4.0 | [69] |

## 6.7   STABILIZATION OF PHYTASE

Enzymes present a low stability during long-term storage and it has been attributed to various factors, as environmental moisture or microbial contamination. Slow changes in the environment conditions, such as pH or temperature, may cause denaturation and the loss of a three-dimensional structure of the enzyme [78]. When denaturation of enzymes is found in a solution, it may be due to hydration of the protein by the formation of hydrogen bounds. The use of polyols, sugars or polymers has been shown to have an effect to stabilize enzymes and preserve their activities [79]. To stabilize an enzyme, it is needed to suppress protein unfolding and retain the catalytic activity. Glycerol has been used as a stabilizer of different enzymes [58, 60, 61]. The mechanism by which proteins are stabilized is not completely understood, however there are some possible explanations. Enzymes which are in aqueous solution might be maintained with polyols that replace the bonds of water and the formation of a water shell around the protein [78]. Denaturalization requires molecular movability of the protein and by means of formation of hydrogen bonds and the creation of a compact state will force the enzyme to keep immobilized and will avoid the unfolding of the protein [80, 81]. There are studies that show the effect of glycerol as a stabilizer in very low temperatures, like –80°C and –20°C, in 13-Hydroperoxide Lyase, in which the result suggests that low concentrations of glycerol are more beneficial for this enzyme [79]. However, Vagenende [82] found that the concentration of glycerol has not a significant effect over stabilization of proteins. This shows that selection of the correct conditions and of appropriate additives depend on the enzyme, as in the case of phytases that have been stabilized with glycerol with concentrations of 25–35% [64, 74].

Sugars have also been used as stabilizers, despite the fact that the way they works is not well understood. Vitrification of sugars explains somehow the way sugars protect proteins. Vitrification is based on the idea of immobilizing the protein in a rigid matrix and a glassy amorphous phase, as it was explained before, the unfolding of proteins needs space, which is limited by the matrix formed by sugars. Aggregation is inhibited in the glassy state which is important to avoid the unfolding of enzymes [83, 84]. Food and Drug Administration (FDA) [85] has approved lyophilized products using sugars like trehalose, sucrose, and lactose as part of the formulation of the drug, mainly antibodies, hormones, and enzymes. One of the aspects to consider in sugar stabilizers is glass transition temperature, which is the temperature where solid-state changes to a rubbery form. When this temperature is overcome, sugars can form vitrification, but it should be taken into consideration that this temperature must not be used for long periods of time. The combination of polyols as glycerol, as well as sugars, may form an interaction and improve the effect over the protein, but this combination needs to be examined [86].

Rodríguez Fernández [73] used glycerol in two concentrations (25 and 35%) to stabilize phytases. Some results showed a solution of phytases in glycerol 35% (v/v) at 4°C with a residual activity of 70% in the 16th week and the linearization of the Arrhenius equation indicates an outcome of 28 weeks half-life.

This suggests the use of polyoles and sugars as promising stabilizers in a wide rank of temperatures and concentrations. The combination with sugars may be an option for the stabilization of phytases.

## 6.8 CONCLUSION

Phytase production has been growing considerably in the last years so the development of better enzymes with more efficient properties has been the subject of interest in recent years and this is the reason why research is taking a different direction towards the recognition of stable and thermotolerant enzymes also looking for its higher production at low-cost through the utilization of agro-industrial wastes. In order to obtain efficient phytases that can be used in animal feeding processes, different steps have taken their place, like: improvement of production and purification methods, implementation of genetic engineering, stabilization of these enzymes with natural materials and optimization of the conditions in fermentation to obtain sufficient phytases ready to be used on animal feeding. The more the phytases' market grows, the bigger the opportunity of finding new ways for improvement and innovation of the production of phytases in the current market.

## KEYWORDS

- Arrhenius equation
- phytase production
- polyoles

## REFERENCES

1. Simons, P. C., Versteegh, H. A., Jongbloed, A. W., Kemme, P. A., Slump, P., Bos, K. D., et al., (1990). Improvement of phosphorus availability by microbial phytase in broilers and pigs. *Br. J. Nutr., 64*(2), 525–540.
2. Lwin, C. M., Murakami, M., & Hashimoto, S., (2017). Resources, conservation and recycling the implications of allocation scenarios for global phosphorus flow from agriculture and wastewater. *Resources, Conserv. Recycl., 122,* 94–105.
3. Shastak, Y., & Rodehutscord, M., (2013). Determination and estimation of phosphorus availability in growing poultry and their historical development. *Worlds. Poult. Sci. J.,* 69(3), 569–586.
4. Menezes-Blackburn, D., Gabler, S., & Greiner, R., (2015). Performance of seven commercial phytases in an in vitro simulation of poultry digestive tract. *J. Agric. Food Chem., 63*(27), 6142–6149.
5. Rodehutscord, M., (1013). Determination of phosphorus availability in poultry. *Worlds. Poult. Sci. J., 69*(2), 525–540.
6. Humer, E., & Zebeli, Q., (2017). Grains in ruminant feeding and potentials to enhance their nutritive and health value by chemical processing. *Anim. Feed Sci. Technol., 226,* 133–151.
7. Jarrett, J. P., Wilson, J. W., Ray, P. P., & Knowlton, K. F., (2014). The effects of forage particle length and exogenous phytase inclusion on phosphorus digestion and absorption in lactating cows. *J. Dairy Sci., 97*(1), 411–418.
8. Amerah, A. M., Ravindran, V., Lentle, R. G., & Thomas, D. G., (2016). Feed particle size: Implications on the digestion and performance of poultry. *Worlds. Poult. Sci. J., 63,* 439–455.
9. Olgun, O., & Aygun, A., (2017). Nutritional factors affecting the breaking strength of bone in laying hens. *Worlds. Poult. Sci. J., 72,* 821–832.

10. Cordell, D., (2010). *The Story of Phosphorus Sustainability Implications of Global Phosphorus Scarcity for Food Security*, Department of Water and Environmental Studies.

11. Udo de Haes, H.A., Smit, W. J., & Van der W., (2009). Phosphate – from Surplus to Shortage. *Minist. Agric. Nat. Food Qual.*, 1–16.

12. Frontela, C., Ros, G., & Martínez, C., (2008). Phytase use as functional ingredient in foods. *Arch. Latinoam. Nutr.*, *58*(3), 215–220.

13. Neira, V. A. A., Vielma, N., Reyna, E. N., Iliná, A., Álvarez, G. M., Gerardo, J., et al., (2013). Fundamental Aspects of phytases. *Investig. y Cienc. la Univ. Autónoma Aguascalientes.*, *21*(57), 58–63.

14. Mikulski, D., & Klosowski, G., (2015). Phytic acid concentration in selected raw materials and analysis of its hydrolysis rate with the use of microbial phytases during the mashing process. *J. Inst. Brew.*, *121*(2), 213–218.

15. Greiner, R., & Konietzny, U., (2006). Phytase for food application. *Food Technol. Biotechnol.*, *44*(2), 125–140.

16. Kumar, V., Sinha, A. K., Makkar, H. P. S., & Becker, K., (2010). Dietary roles of phytate and phytase in human nutrition: A review. *Food Chem.*, *120*(4), 945–959.

17. Bennett, E. M., Carpenter, S. R., & Caraco, N. F., (2001). Human impact on erodable phosphorus and eutrophication: A global perspective. *Bioscience*, *51*(3), 227.

18. Bhavsar, K., & Khire, J. M., (2014). Current research and future perspectives of phytase bioprocessing. *R. Soc. Chem.*, *4*, 26677–26691.

19. Suzuki, U., & Yoshimura, K. T. M., (1907). About the enzyme "phytase," which splits anhydro-oxy-methylene diphosphoric acid. *Bull. Coll. Agric. Tokyo Imp. Univ.*, *7*, 503–512.

20. Li, R., Zhao, J., Sun, C., Lu, W., Guo, C., & Xiao, K., (2010). Biochemical properties, molecular characterizations, functions, and application perspectives of phytases. *Front. Agric. China*, *4*(2), 195–209.

21. Nava, R. E., Álvarez, M., & Hernández, M., (2015). Microencapsulation of bioactive compounds. *Investigación. y Ciencia Universidad Autónoma Aguascalientes.*

22. Konietzny, U., & Greiner, R., (2002). Molecular and catalytic properties of phytate-degrading enzymes (Phytases). *Int. J. Food Sci. Technol.*, *37*, 791–812.

23. Jain, J., & Sapna, S. B., (2016). Characteristics and biotechnological applications of bacterial phytases. *Process Biochem.*, *51*(2), 159–169.

24. Mukhametzyanova, A. D., Akhmetova, A. I., & Sharipova, M. R., (2012). Microorganisms as phytase producers. *Microbiology*, *81*(3), 267–275.

25. Bermudez, M., Mortier, J., Rakers, C., Sydow, D., & Wolber, G., (2016). More than a look into a crystal ball: Protein structure elucidation guided by molecular dynamics simulations. *Drug Discov. Today*, *21*(11), 1799–1805.

26. Oakley, A. J., (2010). The structure of aspergillus niger phytase phya in complex with a phytate mimetic. *Biochem. Biophys. Res. Commun.*, *397*(4), 745–749.

27. Ha, N. C., Kim, Y. O., Oh, T. K., & Oh, B. H., (1999). Preliminary x-ray crystallographic analysis of a novel phytase from a bacillus amyloliquefaciens strain. *Acta Crystallogr. Sect. D Biol. Crystallogr.*, *55*, 691–693.

28. Lerchundi, M. G., (2006). Phytase enzyme obtention from fungal strain of *Aspergillus ficuum* by solid and submerged fermentation. Dissertation., 164.

29. Singh, B., & Satyanarayana, T., (2009). Characterization of a HAP – phytase from a thermophilic mould *sporotrichumthermophile*. *Bioresour. Technol.*, *100*(6), 2046–2051.

30. Zhang, G. Q., Dong, X. F., Wang, Z. H., Zhang, Q., Wang, H. X., & Tong, J. M., (2010). Purification, characterization, and cloning of a novel phytase with low pH optimum and strong proteolysis resistance from *Aspergillus ficuum* NTG-23. *Bioresour. Technol.*, *101*(11), 4125–4131.

31. Rani, R., & Ghosh, S., (2011). Production of phytase under solid-state fermentation using rhizopus oryzae: Novel strain improvement approach and studies on purification and characterization. *Bioresour. Technol.*, *102*(22), 10641–10649.

32. Salmon, D. N. X., Piva, L. C., Binati, R. L., Rodrigues, C., Vandenberghe, L. P. D. S., Soccol, C. R., et al., (2012). A bioprocess for the production of phytase from *schizophyllum commune*: Studies of its optimization, profile of fermentation parameters, characterization and stability. *Bioprocess Biosyst. Eng.*, *35*(7), 1067–1079.

33. Sapna, S. B., (2014). Phytase production by *Aspergillus oryzae* in solid-state fermentation and its applicability in dephytinization of wheat ban. *Appl. Biochem. Biotechnol.*, *173*(7), 1885–1895.

34. Bala, A., Sapna, J. J., Kumari, A., & Singh, B., (2014). Production of an extracellular phytase from a thermophilic mould *Humicola nigrescens* in solid-state fermentation and its application in dephytinization. *Biocatal. Agric. Biotechnol.*, *3*(4), 259–264.

35. Ranjan, B., Singh, B., & Satyanarayana, T., (2015). Characteristics of recombinant phytase (rSt-Phy) of the thermophilic mold *Sporotrichum thermophile* and its applicability in dephytinizing foods. *Appl. Biochem. Biotechnol.*, *177*(8), 1753–1766.

36. Boukhris, I., Farhat-Khemakhem, A., Bouchaala, K., Virolle, M. J., & Chouayekh, H., (2016). Cloning and characterization of the first actinomycete β-propeller phytase from *Streptomyces* Sp. US42. *J. Basic Microbiol.*, *56*(10), 1080–1089.

37. Gaind, S., & Singh, S., (2015). Production, purification and characterization of neutral phytase from thermotolerant *Aspergillus flavus* ITCC 6720. *Int. Biodeterior. Biodegrad.*, *99*, 15–22.

38. Sato, V. S., Jorge, J. A., & Guimarães, L. H. S., (2016). Characterization of a thermotolerant phytase produced by rhizopus *microsporus* var. microsporus biofilm on an inert support using sugarcane bagasse as carbon source. *Appl. Biochem. Biotechnol.*, 1–15.

39. Maurya, A. K., Parashar, D., & Satyanarayana, T., (2017). Bioprocess for the production of recombinant HAP phytase of the thermophilic mold *sporotrichum thermophile* and its structural and biochemical characteristics. *Int. J. Biol. Macromol.*, *94*, 36–44.

40. Yu, P., & Chen, Y., (2013). Purification and Characterization of a Novel Neutral and Heat-Tolerant Phytase from a Newly Isolated Strain *Bacillus nealsonii* ZJ0702. *BMC Biotechnol.*, *13*(1), 1.

41. Kumari, A., Satyanarayana, T., & Singh, B., (2016). Mixed substrate fermentation for enhanced phytase production by thermophilic mould *Sporotrichum thermophile* and its application in beneficiation of poultry feed. *Appl. Biochem. Biotechnol.*, *178*(1), 197–210.

42. Pandey, A., Szakacs, C. R. Soccol, J. A. Rodriguez-Leon, V., & Soccol, T., (2001). Production, purification and properties of microbial phytases. *Bioresour. Technol.*, *77*(3), 203–214.

43. Ebune, A., Al-Asheh, S., & Duvnjak, Z., (1995). Effects of phosphate, surfactants and glucose on phytase production and hydrolysis of phytic acid in canola meal by *Aspergillus ficuum* during solid-state fermentation. *Bioresour. Technol.*, *54*, 241–247.

44. Ebune, A., Al-Asheh, S., & Duvnjak, Z. (1995). Production of phytase during solid-state fermentation using *Aspergillus ficuum* NRRL 3135 in canola meal. *Bioresour. Technol.*, *53*(1), 7–12.

45. Nair, V.C., & Duvnjak, Z., (1990). Reduction of phytic acid content in canola meal by *Aspergillus ficuum* in solid-state fermentation process. *Appl. Microbiol. Biotechnol.*, *34*(2), 183–188.

46. Wang, H. L., Swain, E. W., & Hesseltine, C. W., (1980). Phytase of molds used in oriental food fermentation. *J. Food Sci.*, *45*, 1262–1266.

47. Becerra, M. I. G. S., (1996). Yeast β-galactosidase in solid-state fermentations. *Enzyme Microb. Technol.*, *19*(1), 39–44.

48. Sandhya, A., & Sridevi, P. S. D., (2015). Production and optimization of phytase by *Aspergillus niger*. *Int. J. Pharm. Pharm. Sci.*, *7*(5), 152–157.

49. Kanti, A., (2016). Comparison of *neurospora crassa* and *neurospora sitophila* for phytase production at various fermentation temperatures. *Biodiversitas, J. Biol. Divers.*, *17*(2), 769–775.

50. Shivanna, G. B., & Venkateswaran, G., (2014). Phytase production by *Aspergillus niger* CFR 335 and *Aspergillus ficuum* SGA 01 through Submerged and solid-state fermentation. *The Scientific World Journal*, 392615, 1–6.

51. Joshi, S., & Satyanarayana, T., (2015). Bioprocess for efficient production of recombinant *Pichia anomala* phytase and its applicability in dephytinizing chick feed and whole wheat flat indian breads. *J. Ind. Microbiol. Biotechnol.*, *42*(10), 1389–1400.

52. Singh, N., Kumari, A., Gakhar, S. K., & Singh, B., (2015). Enhanced cost-effective phytase production by *Aspergillus niger* and its applicability in dephytinization of food ingredients. *Microbiol.*, *84*(2), 219–226.

53. Salmon, D. N. X., Fendrich, R. C., Cruz, M. A., Montibeller, V. W., Vandenberghe, L. P. S., Soccol, C. R., et al., (2016). Bioprocess for phytase production by *Ganoderma* Sp. MR-56 in different types of bioreactors through submerged cultivation. *Biochem. Eng. J.*, *114*, 288–297.

54. García-Mantrana, I., Yebra, M. J., Haros, M., & Monedero, V., (2016). Expression of bifidobacterial phytases in *lactobacillus casei* and their application in a food model of whole-grain sourdough bread. *Int. J. Food Microbiol.*, *216*, 18–24.

55. Skowronski, T., (1978). Some properties of partially purified phytase from *Aspergillus niger*. *Acta Microbiológica Polónica*, *27*(1), 41–48.

56. Kujawski, M., & Zyła, K., (1992). Relationship between citric acid production and accumulation of phytate-degrading enzymes in *Aspergillus niger* mycelia. *Acta Microbiológica Polónica*, 41(3–4), 187–191.

57. Zyla, K., & Gogol, D., (2002). *In vitro* efficacies of phosphorolytic enzymes synthesized in mycelial cells of *Aspergillus niger* AbZ4 grown by a liquid surface fermentation. *J. Agric. Food Chem.*, *50*(4), 899–905.

58. Vats, P., & Banerjee, U. C., (2002). Studies on the production of phytase by a newly isolated strain of *Aspergillus niger* var teigham obtained from rotten wood-logs. *Process Biochem.*, 38(2), 211–217.

59. Borjian, B. M., Ranaei, S. S.O., Harati, J., Yousefian, S., & Moradi, S., (2013). Cloning, expression and culture optimization of gene encoding *Aspergillus niger* NRRL3135 phytase in hansenula polymorpha host. *Koomesh, 15*(1), 17–21.

60. Monteiro, P. S., Guimarães, V. M., De Melo, R. R., & De Rezende, S. T., (2015). Isolation of a thermostable acid phytase from *Aspergillus niger* UFV-1 with strong proteolysis resistance. *Braz. J. Microbiol., 46*(1), 251–260.

61. Shi, C., He, J., Yu, J., Yu, B., Mao, X., Zheng, P., Huang, Z., & Chen, D., (2016). Physicochemical properties analysis and secretome of *Aspergillus niger* in fermented rapeseed meal. *PLoS One, 11*(4).

62. Chi-Wei Lan, J., Chang, C. K., & Wu, H. S., (2014). Efficient production of mutant phytase (phyA-7) derived from *Selenomonas ruminantium* using recombinant *Escherichia coli* in pilot scale. *J. Biosci. Bioeng., 118*(3), 305–310.

63. Costa, M., Torres, M., Magariños, H., & Reyes, A., (2010). Production and partial purification of *Aspergillus ficuum* hydrolytic enzymes in solid-state fermentation of agroindustrial residues. *Colomb. Biotecnol., 12*(2), 163–175.

64. Ma, X., & Shan, A., (2002). Effect of germination and heating on phytase activity in cereal seeds. *J. Anim. Sci., 15*(7), 1036–1039.

65. Nava, E. R., (2014). Development of technology for functional additives production based on microencapsulated of phytases and probiotics for monogastric animals. PhD Dissertation, Autonomus University of Coahuila.

66. Neira, V. A., (2013). Phytases production with native strains of Aspergillus spp. by solid-state fermentation of agroindustrial wastes. Master Dissertation, Autonomous University of Coahuila.

67. Awad, G. E. A., Helal, M. M. I., Danial, E. N., & Esawy, M. A., (2014). Optimization of phytase production by *Penicillium purpurogenum* GE1 under solid-state fermentation by using box-behnken design. *Saudi J. Biol. Sci., 21*(1), 81–88.

68. Abid, N., Khatoon, A., Maqbool, A., Irfan, M., Bashir, A., Asif, I., et al., (2016). Transgenic expression of phytase in wheat endosperm increases bioavailability of iron and zinc in grains. *Transgenic Res., 26*(1), 109–122.

69. Gupta, R. K., Gangoliya, S. S., & Singh, N. K., (2013). Reduction of phytic acid and enhancement of bioavailable micronutrients in food grains. *J. Food Sci. Technol., 52*(2), 676–684.

70. Hao, T., Xiang, W., Liyuan, X., & Zhongqian-Huang, B. G., (2015). Cloning, overexpression, and characterization of a metagenome-derived phytase with optimal activity at low pH, *25*(6), 930–935.

71. Erpel, F., Restovic, F., & Arce-Johnson, P., (2016). Development of phytase-expressing *Chlamydomonas reinhardtii* for monogastric animal nutrition. *BMC Biotechnol., 16*(1), 29.

72. Becker, E. W., (2007). Micro algae as a source of protein. *Biotechnol Adv., 25*, 207–210.

73. Rodríguez-Fernández, D. E., Parada, J. L., Medeiros, A. B. P., Carvalho, J. C., De Lacerda, L. G., Rodríguez-León, J. A., & Soccol, C. R., (2013). Concentration by ultrafiltration and stabilization of phytase produced by solid-state fermentation. *Process Biochem.*, 13–15.

74. Thomas, C. R., & Geer, D., (2010). Effects of shear on proteins in solution. *Biotechnol. Lett.*, *33*(3), 443–456.

75. Escobin-M Opera, L., Ohtani, M., Sekiguchi, S., Sone, T., Abe, A., Tanaka, M., Meevootisom, V., & Asano, K., (2012). Purification and characterization of phytase from *Klebsiella pneumoniae* 9-3B. *JBIOSC*, *113*(5), 562–567.

76. Wodzinski, R. J. U. A., (1996). Phytase. *Adv. Appl. Microbiol.*, *42*, 263–302.

77. Graminho, E. R., Takaya, N., Nakamura, A., & Hoshino, T., (2015). Purification, biochemical characterization, and genetic cloning of the phytase produced by *Burkholderia* Sp. strain a13. *Appl. Microbiol. Biotechnol.*, *23*, 15–23.

78. Costa, S. A., Tzanov, T., Filipa, C. A., Paar, A., Gübitz, G. M., & Cavaco-Paulo, A., (2001). Studies of stabilization of native catalase using additives. *Enzyme Microb. Technol.*, *30*(3), 387–391.

79. Jacopini, S., Vincenti, S., Mariani, M., Brunini-Bronzini de Caraffa, V., Gambotti, C., Desjobert, J. M., et al., (2016). Activation and stabilization of olive recombinant 13-hydroperoxide lyase using selected additives. *Appl. Biochem. Biotechnol.*, *182*(3), 1000–1013.

80. Pazhang, M., Mehrnejad, F., Pazhang, Y., Falahati, H., & Chaparzadeh, N., (2016). Effect of sorbitol and glycerol on the stability of trypsin and difference between their stabilization effects in the various solvents. *Biotechnol. Appl. Biochem.*, *63*(2), 206–213.

81. Khajehzadeh, M., Mehrnejad, F., Pazhang, M., & Doustdar, F., (2016). Effects of sorbitol and glycerol on the structure, dynamics, and stability of *Mycobacterium tuberculosis* pyrazinamidase. *Int. J. Mycobacteriology*, *5*, 138–139.

82. Vagenende, V., Yap, M. G. S., & Trout, B. L., (2009). Mechanisms of protein stabilization and prevention of protein aggregation by glycerol. *Biochemistry*, *48*(46), 11084–11096.

83. Neira, V. A. A., (2017). Phytases production by *Aspergillus niger* in pilot scale for the formulation of a product as functional additive for the diet of non-ruminant's animals. PhD Dissertation, Autonomous University of Coahuila, Federal University of Pernambuco.

84. Mensink, M. A., Frijlink, H. W., Van der Voort Maarschalk, K., & Hinrichs, W. L. J., (2017). How sugars protect proteins in the solid-state and during drying (review): Mechanisms of stabilization in relation to stress conditions. *Eur. J. Pharm. Biopharm.*, *114*, 288–295.

85. American Food and Drug Administration, FDA approved drug products http://www. accessdata.fda.gov/scripts/cder/daf/index.cfm.

86. Crowe, L. M., (2002). Lessons from nature: The role of sugars in anhydrobiosis. *Comp. Biochem. Physiol. – A Mol. Integr. Physiol.*, *131*(3), 505–513.

# CHAPTER 7

# PHYTOCHEMICAL MOLECULES FROM FOOD WASTE AND DESERT PLANTS FOR CONTROL OF FOODBORNE PATHOGEN BACTERIA

ROSARIO ESTRADA-MENDOZA,[1]
ADRIANA C. FLORES-GALLEGOS,[1] JUAN A. ASCACIO-VALDES,[1]
CRISTÓBAL N. AGUILAR,[1] SANDRA C. ESPARZA-GONZÁLEZ,[2]
SANDRA L. CASTILLO-HERNÁNDEZ,[3] and
RAUL RODRÍGUEZ-HERRERA[1]

[1] Food Research Department, School of Chemistry, Autonomous University of Coahuila, Boulevard Venustiano Carranza and José Cárdenas s/n, República Oriente, Saltillo 25280, Coahuila, México, E-mail: raul.rodriguez@uadec.edu.mx

[2] School of Medicine, Autonomous University of Coahuila, Saltillo, Coahuila, México

[3] School of Biology, Autonomous University of Nuevo Leon, Monterrey Nuevo León, México

## ABSTRACT

Currently, with the constant increase of global commerce, the possibility of propagation of foodborne pathogen bacteria has arisen. This may generate animal health problems in both, developing and developed countries. Synthetic chemicals have been used for control of pathogenic bacteria. However, now, customers demand low of null presence of synthetic chemicals in food. For these reasons, the food industry is searching for bioactive molecules with capacity to control pathogens from natural sources. Use of different plant extracts with antimicrobial activity has been reported.

These extracts are a sustainable and profitable possibility for foodborne pathogen bacteria control. In this chapter, different applications of phytochemical molecules from food waste and desert plants to control foodborne pathogen bacteria are discussed.

## 7.1 INTRODUCTION

Foodborne diseases (FBD) are public health problem at global level that affects millions of people every year [1]. The foodborne diseases cause morbidity and mortality at different proportions in different countries [2]. But, there are large differences between developed and under-developing countries [3]. The World Health Organization (WHO) defines FBD as "disease caused by infectious agents (bacteria, viruses or parasites) or non-infectious (inadvertent toxins produced by microorganisms and toxic chemicals) that enter the body through ingestion of contaminated food or water" [2, 4]. These pathogen agents are capable of causing minor damage or life-threatening [5]. Although, food security has generally improved, progress is uneven and outbreaks of food poisoning or infection are frequent in several countries [6]. For both, consumers and the food industry, microbial food safety is a major concern [7].

Symptoms of FBD include: stomach pain, vomiting, nausea, diarrhea, which typically last for two to three days in almost all individuals. The damage is in most of its forms harmless but, in other cases (certain patients) may have severe complications. Among the severest damage are: stillbirths (for instance, infections with *Listeria*), hospitalization due to sepsis, nerve paralysis, and arthritis [5], hepatic, and renal (infections with *Escherichia coli*) syndromes [4]. Foods have the latent threat of contamination (with different infectious and/or non-infectious agents) at any stage of the process, from production to consumption. The risk of contamination is maintained throughout the production chain, starting with possible environmental contamination (water, soil or air), the production stage, to the consumption stage [8]. Some common ways to spread pathogen microorganisms are via the fecal-oral route, fruits, and vegetables irrigated with contaminated water (with microorganisms such as *Shigella*), or when food handlers have skin lesions caused by an organism such as *Staphylococcus aureus*. Other via of contamination may be cases in which this type of microorganism can stay in a healthy animal's gastrointestinal tract and during slaughter step, the microorganism that is infecting and contaminating that animal could contaminate the meat, if it is not cooked in the best way [5].

Pests lurk and attack food crops around the world, especially those of fungal origin [9]. This represents, in its least serious effect: economic losses during and after harvest (approximately 50%), but in health aspects, the effect of fungal or bacterial intoxications can cause diseases in plants, animals, and humans. For example, some species of the *Fusarium, Aspergillus, Penicillium,* and *Alternaria* genera produce mycotoxins, highly dangerous compounds, carcinogenic, mutagenic, teratogenic, immunosuppressive or even deadly for human beings [10]. Consumers will be in danger of consuming contaminated foods or their derivatives, even the danger remains in products already processed [11]. On the other hand, in order to avoid microbiological contaminations, synthetic chemical substances have been used in crops, but abuse of these antimicrobial synthetic chemicals (inhibitors of biosynthesis of sterols, aromatic hydrocarbons, benzimidazoles, imazalil, sodium orphenylphenate and thiabendazole, pyrimethanil (anilinopyrimidine) and fludioxonil (phenylpyrrole)) have led to emerging microbial populations with resistance, contaminant retention in soil and water, and crops with high residual contents of these products that have a toxic effect on humans through the whole food chain [10, 12].

An outbreak is not only restricted to the local area where it happened, the current world is connected and it makes that even a local outbreak is a viable threat at the global level. Factors such as migration, travel, and international trade have breached any "country borders," making possible spread of dangerous microorganisms causing foodborne infections, which can affect people health in different geographical regions, including places far from the point of origin [13]. Other forms of foods contamination can occur by improperly cleaned of cooking utensils and/ or contaminated kitchen. In addition, when food is kept at inadequate temperatures, organism such as *Clostridium* species and *Bacillus cereus* can produce toxins or spores that when ingested, germinate in the human gut when conditions for microorganism growth are more favorable [5]. FBD cause economic losses to production chain, tourism, together with social, cultural, and health aspects, this type of infections affect the progress of nations, and not only that, but indirectly affecting the objectives at the international level of various organizations that try to eradicate health and poverty problems [13].

## 7.1.1   SPOILER MICROORGANISMS

Change process that a food undergoes making it unacceptable for consumption, is defined as food decomposition. This process may be the result of food enzymatic activity, however, it is the activity of its microflora that is the main cause of deterioration [14]. Some non-pathogenic microorganisms cause quality detriment in a food system producing unpleasant flavors and odors, damaging food texture, as well as decreasing or changing color, which inevitably leads to customer rejection of the product [15]. The changes perceived in a food are the result of the metabolic activity of microorganisms (fungi, yeasts, and bacteria). These characteristics can be measured by physical parameters (number of different metabolites, gases, color, pH) and sensory analysis [16].

Generally, molds, and yeasts tend to affect the food under aqueous activity and a low pH. Deterioration is evidenced by an evident growth on food surface, as well as, fermentation of sugars which is perceived by the excessive of gas production ($CO_2$). The fungi associated to fruits and vegetables post-harvest are: *Aspergillus, Mucor, Penicillium,* and *Rhizopus*. Yeasts attacking syrups, sauces, soda, and dressings are mainly belonging to *Torulaspora, Candida, Lachancea, Saccharomyces,* and *Zygosaccharomyces* [17]. The ability of yeasts and molds to contaminate soft drinks and other niches lies in their ability to survive in acid and carbonate contexts. In beverages, their presence is evidenced with abundant production of carbon dioxide ($CO_2$), excessive effervescence, turbidity, decrease of Brix degrees. Refreshment beverages frequently include *Kloeckera apiculata, Pichia membranifaciens, Zygosaccharomyces bailii, Zygosaccharomyce srouxii, Zygosaccharomyces bisporus, Schizosaccharomyces pombe* and *Saccharomyces exiguus,* which are associated with food spoilage [18].

Animal products are susceptible to Gram-positive bacteria (mainly *carbon dioxide*) [19], lactic acid bacteria (*Lactococci, Lactobacilli, Leuconostoc, Weissella,* and *Carnobacteria*) and *Enterococci* which are of the most prominent spoiler bacterial species. The famous *Brochothrix thermosphacta,* which deteriorates various food matrices, but is mainly, detected in meat [20]. The *Lactobacillus alimentarius, Lactobacillus sakei* and *Lactobacillus curvatus* genera are associated with deterioration of marine products and *Lactobacillus sakei* deteriorating vacuum-packed poultry. Likewise, the *Carnobacterium divergens* and *Carobacterium maltaromaticum* genera attack meat and seafood products that were bottled under vacuum and in controlled atmospheres, causing the "blown effect" (production and storage of gas). Other gas-producing bacteria are *Enterobacteriaceae* (*Enterococcus faecalis* and *Enterococcus faecium*), responsible for metabolizing amino acids and causing a very unpleasant odor. *Clostridium algidixylanolyticum, Clostridium frigidicarnis, Clostridium frigoris, Closdridium gasigenes* and *Clostridium algidicarnis*) and *Leuconostoc gasicomitatum*are related to acrid and buttery acrid odors. *Leuconostoc* is the main detractor of vegetable quality [16].

Within the Gram-negative spoilers have been reported *Serratia, Hafnia, Pseudomonas, Shewanella baltica* and *Photobacterium phosphore* (the latter for cold water products being a psychophilous, $CO_2$ resistant and halophilic bacterium). *Pseudomonas spp* predominate in raw foods stored under aerobic conditions, particularly *Pseudomonas fluorescens, Pseudomonas putida, Pseudomonas fragi, Pseudomonas gessardii, Pseudomonas lundensis* and *Pseudomonas fluorescenslik*, the last four being related to dairy products (still pasteurized). *Pseudomonas chlororaphis* is a spoiler for vegetable products such as seeds, onion, carrot, potatoes, and lettuce. Spore-forming bacteria of the *Clostridium* and *Bacillus* genera (specifically *Bacillus amyloliquefaciens, Bacillus subtilis* and *Bacillus pumilusson)* are related to deterioration of various raw materials of bakery products such as eggs, cheese, and chocolate products [21]. Microorganisms responsible for decomposition can form a biofilm, defined as an aggregation of microorganisms growing harmoniously on a surface affected by different factors such as: material surface properties, specific bacterial strain and environmental parameters (pH, temperature, and nutrients) [22]. The major danger with biofilms is the association of spoilages with pathogenic microorganisms intended to survive in the inhospitable environment. This is important in many foodborne infections [8].

## 7.1.2   PATHOGEN MICROORGANISMS

It is difficult to estimate accurately the global incidence of foodborne diseases; the challenge is to have a real measuring of foodborne diseases, the problems for that, only a little fraction of people sick for contaminated food look for medical care [6]. Few cases of foodborne diseases are treated adequately, reported to public health authorities and recorded in disease statistics, and the causal link between certain chronic diseases (such as kidney or liver failure or cancer) result from consumption of contaminated food such appear long after the ingestion and never are related to each other [23]. Assessment and impact of an outbreak of foodborne disease varies according to the pathogen or toxin responsible, volume of contaminated food, step of the production chain where the contamination existed, geographical situation and number of people who consumed the contaminated food [24]. Although, there are several pathogenic agent that cause food illness, etiology of foodborne diseases are mostly of bacterial origin [5]. Four bacteria top the list of causative agents of food infections: *Salmonella* spp., *Campylobacter* spp., *Escherichia coli*, and in recent years, *Listeria monocytogenes* has been added because of important occurrences in different outbreaks [13].

## 7.1.2.1   SALMONELLA SPP

Salmonellosis is the most frequently reported food infection, according to WHO data and the large outbreaks reported in recent times, and chicken and meat being the main transmission vehicles [3]. In order to stop *Salmonella* infection, physical (sterilization, pasteurization), chemicals (chemical products) and microbiological treatments have been used on the final food products, and even as a preventive way, use of vaccines made with attenuated *Salmonella* bacteria cells had been used [3].

## 7.1.2.2   CAMPYLOBACTER SPP

These bacteria are found in natural environment (water surfaces) and are part of the natural intestinal microbiota of a wide range of wild and domestic birds (poultry), in addition to pigs and cattle, lodging in the intestinal tract, although, it is also capable to survive for long periods of time outside a given host [25]. Transmission to humans occurs during the slaughter of animals and further processing. Besides, to the symptoms of gastrointestinal infection, there is a risk of post-infectious squeals and aggravation or chronic complications, such as Guillain-Barré syndrome, Miller-Fisher syndrome, Bickerstaff encephalitis and development of inflammatory bowel diseases (Crohn's disease, ulcerative colitis and irritable bowel syndrome) [26]. More specifically, *Campylobacter jejuni* is the main cause of diarrheal diseases. Virulence of this bacterium is related to formation of biofilms, making it a bacterium resistant to environmental treatments and antibiotics. Formation of biofilm is result of expression of fla*A* and fla*B* genes, which allow this pathogen to respond to different environments (different food matrices and adhesion to surfaces) and make it a concern worldwide [27].

## 7.1.2.3   *LISTERIA MONOCYTOGENES*

It is widely distributed in the environment and animals and can enter the human gastrointestinal system through foods such as dairy products, including soft cheeses, and in raw and undercooked meat, poultry, seafood, and related products [28]. *Listeria monocytogenes* in recent decades has been recognized as an emerging pathogen transmitted by food and is a worldwide concern to the food industry and health authorities. The main symptoms of Listeriosis are a mild and transient febrile illness, but may be invasive with more severe symptoms [29] causing meningitis, sepsis or abortion, but in practice, only pregnant women and people with immunological defects are in danger of infection [28]. This bacterium commonly inhabits gastrointestinal tract of mammals and birds, however, various strains are associated with diarrheal disorders and others even fatal diseases. Infections outside the gastrointestinal tract are the cause of urinary tract infections.

## 7.1.2.4   *ESCHERICHIA COLI*

Different pathogenic strains of this bacterium are grouped into enterotoxigenic *E. coli* (ETEC), which produces enterotoxins capable of stimulating the small intestine to secrete electrolytes and water, resulting in diarrhea, causing "traveler's diarrhea"; *E. coli* enteroinvasive, similar to *Shigella*, invade, and damage cells of the colonic mucosa and cause diarrhea disenteric in humans; enteropathogenic *E. coli* (EPEC), which are able to adhere to intestinal cells causing stroke injuries and causes profuse watery infant diarrhea; *E. coli* enteroaggregative (EAEC), which cause aqueous diarrhea by adhesion to the epithelium of the ileum, causing profound damage; Shiga toxin-producing *E. coli* (Stx), which are cytotoxins that inhibit protein synthesis within eukaryotic cells and cause hemorrhagic

(HC) colitis and hemolytic uremic syndrome. Infections spread by the oral-fecal route [30].

### 7.1.3 MICROORGANISMS RESISTANT TO ANTIBIOTICS AND EMERGING MICROORGANISMS

Foodborne diseases began with ancient man [2], and they are as antique, and similar as humankind, nevertheless, affecting all levels of society. But, it is important to note that they are changing over time [31]. The known pathogens appear to acquire new attributes or appear in unexpected foods, appear new and the difficulties to face them are still in force, in addition, population increase sensitized to more microbes, which can increase the risk of suffering from opportunistic pathogens that colonize fresh food [13]. New food consumption trends, new perspective about foodborne illnesses and emerging bacterial pathogens are some of the causes for current food-borne diseases. During outbreaks of food borne diseases, new bacterial pathogens are identified [32], for example, *Cyclospora cayetanensis*, a previously unknown coccidian, emerge as foodborne pathogen [5].

In Wisconsin during 2008, a first outbreak unleashed caused by *Arcobacter butzleri* was observed after ingestion of contaminated chicken. Symptoms produced by this pathogen are similar to those produced by other pathogens such as *Campylobacter.* In spite of *Arcobacter* spp. is difficult to isolate using traditional microbiological methods, it has been isolated from different foods of animal origin and fecal samples of dairy animals, the same way, in production sites from artisanal to an industrial cheese factory [32]. In the same year, in Bangalore (India) an outbreak of avian influenza resulted in import restrictions from this country to the Middle East, causing serious economic losses. Examples such as these mentioned before, help to understand the wide incidence of *Salmonella enteritidis* (the major serotype in most countries) and multi-resistant antibiotics strains of *Salmonella typhimurium* [13].

In 1991, an epidemic of cholera that began in Peru's polluted waters ended with 4000 deaths distributed in various Latin American countries, and for 1998 in E.U.U. were reported outbreaks of *Escherichia coli O157: H7* ready-to-cook pizzas and fresh vegetables, as well as *Clostridium* poisoning pro-foods canned, along with increased incidences of *Campylobacter spp.*, which commonly causes sporadic infections [13]. Another example, *Helicobacter pullorum* is other emerging microorganism founded in the intestine of broiler chickens, laying hens, as in patients with diarrhea, gastroenteritis, inflammatory bowel disease, hepatobiliary, therefore, this microorganism is classified as pathogen with zoonotic potential, but, it is not all, there is very little knowledge about *Helicobacter pullorum* front the host immune system and regarding about the phagocytic response [32].

Significant numbers of sporadic cases increased in last decades; also, it has showed outbreaks caused by other serotypes non-O157 STEC with unusual virulence traits, at the same than *E. coli O104:H4*, Shiga toxin-producing strains have been reported, especially cases with highly pathogenic combination of virulence factors (outbreaks in Germany and Francie 2011). Shiga toxin-producing *Escherichia coli* (STEC) is a group of enteric pathogens cause hemorrhagic colitis and acute kidney failure and chronic post-infection sequelae or in the worst-case death. Further, the apparition of non-O157 STEC serotypes has become a big challenge for public health, regulatory authorities and agro-food sector [32]. The foodborne pathogen *Bacillus cereus* produces heat-resistant spores and is an important hazard for heat-treated, and non-sterilized, foods. In addition, some *Bacillus cereus* strains are psychotropic and may grow in refrigerated foods. The cold adaptation of *Bacillus cereus* makes it one of the few pathogens able to survive during food processing and to multiply in cooked-chilled foods [17]. On the other hand, the acid tolerance response of *Salmonella typhimurium* can be defined as the induced resistance to normally lethal low pH following growth at moderately low pH, or following short-term exposure to mild acid conditions [33]. Due to the ample range of adaptive mechanisms that possess the microorganisms, they are able to respond to environmental stressful, by temporarily changing their patterns of gene expression and/ or modulating the cellular structure or conformation, among those mechanisms are adaptive constitutive or transient mutations [17, 33].

The proliferation of bacterial populations resistant to different antibiotics and appearance of new pathogenic bacteria make imperative the search of new molecules with bactericidal power which should be cheap and easy and rapid obtained [34]. Therefore, food researchers are looking for alternatives in terms of natural preservatives that, in addition to prolonging the shelf life of foods, also ensure their safety. According to the mentioned before, natural molecules from plants are the most besieged today. Natural plant-derived conservers are very interesting compounds because in addition to replacing synthetic preservatives in current use [35], also would cover the growing preference today's consumer that focus on more natural foods [36]. In addition, consumers ask more for food minimally processed and foods with low or nothing (if possible) synthetic additives. Certainly, synthetic antimicrobial products are approved in many countries [7], but today the trend towards natural preservatives is increasing because some adverse health effects of synthetic preservatives have been mentioned [37], and because new forms of alimentation [38]. The food industry is working hard against presence of pathogens and their detection in food, since their presence could cause disastrous economical and health effects [6].

It is known that various plants and/or extracts widely used in the food industry as flavorings [39]. Because of consumer demands for "green" or "natural" foods, the food industry has been inclined to incorporate substances of plant origin that in addition to the aroma properties also offer other benefits such as antioxidant power and/or antibacterial power [15]. Antioxidant compounds and food preserves (both natural and synthetic) are used to prevent oxidation reactions in food and prolong quality and nutritional status (the first case) [36] and to retard and/or prevent the growth of microorganisms (pathogens and spoilers) that degrade food and can because illness (the second case) [18]. Examples of synthetic antioxidants commonly found in food preservation are propyl gallate (PG), butylated hydroxytoluene (BHT), tert-butyl hydroquinone (TBHQ) and butylated hydroxyanisole (BHA). However, some reports indicate that BHA and BHT compounds can be harmful [18, 40], and among the most ancient are: acetic acids, propionic acid, benzoic acids, sorbic acid, and their sodium, potassium, and calcium salts. Plant extracts with antimicrobial activity may inhibit presence of food pathogens, which is of great interest for food industry being these plant extracts very popular because are used to delay reactions of oxidative degradation, preserve, and improve nutritional value and prevent microbial deterioration [41].

## 7.2   PHYTOCHEMICALS: A GIFT FROM NATURE

Foods are complex matrixes and their chemical and physical properties most of the time may possible proliferation of many and different microorganisms. Aspects such as limited nutrient availability, osmolality, pH, extreme temperatures and oxidation, are some intrinsic conditions own to each food that avoid or allow pathogen bacteria growing [33]. Herbal plant compounds are an alternative to treating foodborne pathogens and therefore promoting reduction of antibiotic use, that recently have been prohibited by European Union (EU) to use in animal feed. USA is concerned about development of antibiotic resistance of foodborne pathogens, which will cause major problems for animals and humans. Some natural chemicals from plants have antimicrobial activities, which have fewer side effects and are safe for humans and animals [4]. Used as vegetal extracts or as standardized pure compounds, these plant products offer ample opportunities for control of pathogens because of their diverse chemistry [7, 22]. The ability of a phytochemical compound to inhibit or retard microbial growth is determined by the chemical characteristics of phytocompounds. Therefore, properties such as polarity, hydrophobicity/lipophilicity, point of volatility, solubility, pH (to name a few) of molecules are determinant to establish their antimicrobial level for its use in food. In addition, it is also important to know the specific characteristics of the food matrix to be preserved, as well as its storage indices (temperature, humidity), to define the efficient concentration of substance to meet the objectives, also combination of antimicrobials to potentiate their effects [7].

Since ancient times, plants have been used by humans for a wide range of purposes. Use in food (human and animal), to enhance the flavor, color, and aroma [42], fuels, poisons, medications, ornamental, recreational, pollution indicators or currently as bioreactors (genetic modification) to obtain different products [43]. However, it is undoubtedly that to serve as a source of food and pharmacological activities are its main benefits. From the known plants, approximately 5,000 species are intended for food, and of these, only 30 species are sufficient to cover 95% of the world's calorie consumption with three (maize, wheat, and rice) being the most used [43]. For thousands of years, products extracted from plants have been used to treat various diseases and therefore human interest on phytochemical molecules remains valid. According to the World Health Organization, 80% of the world's population undergoes traditional medicine treatments as primary care, mainly using plant extracts and their bioactive compounds [44]. Humanity has relied on use of natural products to extract precursor molecules of new drugs [45]. These compounds are generally non-toxic, many show efficiency even at low concentrations, are economical and environmentally friendly [46].

Knowledge and study of various plant species and their extracts are important for food conservation since presence of different phytochemicals in these plants have a positive and defined action on microbial control [47] and natural sensory of food. Attempts have been made to use crude extracts or isolated phytocompounds as natural preservatives [48]. It is estimated that there are 250,000 different plant species on Earth, which are potential sources of secondary metabolites that may have application as antimicrobials [35]. Therefore, it is necessary to look for new alternatives for specific objectives such as prevention of food pathogens [41].

In Mexico, the ingrained tradition of using plants as the first defense tool for various diseases (infectious or not) has been well documented since ancient times, in manuscripts such as " the Codex Badiano," dating from 1552. In Mexico, the commercialization of a diverse variety of herbal formulas (infusions, drops applications, syrups) and creams or even steam baths called "temazcales" to promote use and benefit of plants [49], nonetheless, there is no standardization of herbal formula ingredients, thus, the knowledge necessary for identification of plants and their precise effects on prokaryotes (bactericidal properties) and eukaryotes (possible toxic damages) are very important [40].

## 7.2.1  DESERT ECOSYSTEM

The Convention on Biological Diversity (CBD) defined the biological diversity of plants as "the variability that exists among living plants in all terrestrial, marine, and aquatic ecosystems." That is, plant biodiversity encompasses all levels of biological organization (from genes to communities) [43]. Excluding wide earth areas where plants have not yet been explored, scientists estimate a total of 350,000 different plants in the world. The most extensive terrestrial biome corresponds to arid lands, which cover more than a third of the earth's surface (about 22% of the world's land surface) [49]. Despite of the large land area in the world and its wide variety of flora and fauna, only about 1% of the world's primary productivity occurs in deserts [50].

In the Northern Hemisphere, the succession of mid-latitude deserts consists of: (1) the Mojave Desert, (2) the Sonora Desert, (3) the Chihuahua Desert and (4) the Great Basin. The Sonoran Desert is the hottest and most humid, and probably the most prosperous in terms of vegetation compared to other deserts in the world [51]. The exploration and exploitation of the great diversity of plants is incipient compared to its estimated plant species amount. There is a considerable opportunity for further domestication and promotion of underutilized crops, such as the desert plant species [43]. It has been mentioned that production of plant secondary metabolites depends to a large extent on climatic conditions, type of soil in which these plants grow and conditions under they are grown [52]. A perfect example for this are plants grown in hostile environments, which produce a greater number of phytocompounds developed as defense against these conditions. In the Mexican Northern part, especially in the semi-arid regions, there are different plants that produce compounds of interest, such as antimicrobials and antioxidants [41].

## 7.2.2   DESERT PLANTS

Mexico is among the five countries highlighted by its wide variety of plants. It has 25,000 registered species and 30,000 have not yet been described [46], it has 32 native genera that cover 591 species, equivalent to 13.55% of the genera and 8.23% of the world's distribution families with the highest number of species showing antibacterial effects which include: *Asteraceae* (57 species). Mexico is home of 314 to 387 genera ranging from 2,000 to 3,000 different species out of a total of 25,000 species grouped into 1,500 genera of the family *Asteraceae, Fabaceae* (28 species), is one of the most extensive and important families from the economic point of view. It has 740 genera with 19,400 species worldwide. Mexico has 16.66% of this number of species which are widely used for its high content of amino acids and fiber and, also been used for medicinal purposes (due to presence of phenolic compounds, saponins, and terpenoids), and other important family is *Lamiaceae* (21 species). This family has more than 3000 species of plants distributed in warm and temperate regions of the world [53].

Plants from the semi-arid regions of Mexico used in traditional medicine against infectious diseases (bacterial nature principally, but viruses and fungi too) are given in the following subsections.

## 7.2.2.1 FLOURENSIA CERNUA (TARBUSH)

Belonging to the *Asteraceae* family, the *Flourensia* genus distinguishes itself by its adaptation to arid or desert areas [54]. It is characterized by its higher production of metabolic substances under stress conditions (drought, high, and/or low temperatures, intense UV radiation) [55]. This genus has 40 species that are characterized by being shrubs with penetrating aroma and resinous leaves with large yellow flowers, which are used to obtain infusions (Figure 7.1) [56]. Nine of the species reported [57] for this genus are native of the North of Mexico, however, this shrub is commonly underutilized in the soils of the Chihuahua desert, where it grows in an area of 35 million ha in the clay soils of the Chihuahuan desert [58]. *Flourensia cernua* D.C. (called colloquially "hojasen," tarbush, blackbrush, varnish bush) *Flourensia retinophylla* and *Flourensia microphylla* the most studied [59]. Plants from this genus produce diverse phytochemicals that have an important biological activity. Among these compounds are: chrysin, apigenin, galangin, kaempferol, sesquiterpenes, jaletin, flavanones, and acetylbenzofurans. From *Flourensia cernua* have been reported flavonoids, acetylenes, p-acetophenones, sesquiterpenoids, dehydroflurenic acid, benzofurans, and benzopyrans [59, 60]. Antifungal activity (*in vitro*) of extract from this plant has been reported against *Rhizoctonia solani, Pythium spp., Fusarium oxysporum* and *Colletotrichum accutatum,* also aqueous extracts have also been tested against *Rhizopus stolonifer, Botrytis cinerea* and *Colletotrichum gloeosporioides* [61–63]. In addition, extracts from this plant are reported as termiticide. The biological activity of these plants has been linked to the presence of benzofurans and benzopyrans, as well as having the ability to form alkyl [63].

## 7.2.2.2 JATROPHA DIOICA (DRAGON'S BLOOD)

This plant belongs to the *Euphorbiaceae* family [64], which has approximately 300 genera and 9,000 species, including large woody trees, climbing lianas to simple weeds. These plant species grow in a diversity of tropical in seasonally dry and subtropical climates. *Jatropha dioica* is indigenous to Mexico and the state of Texas, United States (US) [65]. Mexico has between 43 to 50 genera of this family and 782 to 826 species

**FIGURE 7.1** *Fluorensia cernua.*

being 56% of them endemic of this country [49] *Jatropha dioica* is indigenous to Mexico and the state of Texas, US and exhibits a colorless juice that becomes dark when in contact with the air, hence its common name "dragon blood" [65], although this characteristic is not specific to this specie, as other plant species of different plant genera such as (*Dracaenaceae*), Daemonorops (*Palmaceae*), Croton (*Euphorbiaceae*) and Pterocarpus (*Fabaceae*) also exhibit this resin [66]. Although various species of the genus are used as an ornament, many of them are used as auxiliaries in wound treatment, skin inflammations, rheumatism, and snake bites [64]. Therapeutic effects are associated with the presence of citlalitrione, jatrophona, and riolozatrione (diterpenes), ellagic acid, oxalic acid, a sterol [65] and due to its astringent effect, it is used as antidiarrheal and hemostatic, too [66]. Extracts of *Jatropha dioica* showed biological activity against some important fungi such as *Penicillium purpurogenum, Fusarium spp., A. alternata, R. solani* and *Aspergillus flavus* [67].

### 7.2.2.3   EUCALYPTUS CAMALDULENSIS (EUCALYPTUS)

This plant belongs to the *Myrtaceae* family. Although E*ucalyptus camaldulensis* originates from Australia it is widely adaptable to different climatic regions around the world (Figure 7.2). The main compounds in *Eucalyptus* are phenolic compounds such as gallic acid, protocatechidic acid, ellagic acid, quercetin, quercetin glycol, naringenin, catechin, epicatechin, rutin, quercitrin, apigenin, and myricetin. Leaves and bark have been used around the world as adjuncts to skin diseases, dysentery, malaria, and bacterial diseases (against symptoms such as fever) [68]. The essential oils obtained from *Eucalyptus teretecornis* showed anti-microbial activity against the fungus *Alternaria alternata*, the active fractions correspond to oxygenated terpenoids (β-fenchol and α-eudesmol) [23].

**FIGURE 7.2**   *Eucalyptus camaldulensis*. (A) Full view; (B) Detail of leaves and stem.

## 7.2.2.4  *LARREA TRIDENTATE (CREOSOTE BUSH)*

Larrea is a shrubby plant belonging to the *Zygophyllaceae* family [69]. Common shrub in the Northern region of Mexico, where is called "chaparral" or "gobernadora" (Figure 7.3) [70]. Extracts of this plant are used as drinkable infusions for urinary and infectious disorders and as vaginal showers in gynecological treatments, and treatment against tuberculosis. At low concentrations does not show toxicity, however, high doses cause renal and hepatotoxic damage in humans [70]. Its resin is distinguished by its phytochemical properties.

It has been reported that contain nordihydroguaiaretic acid (NDGA) as the main phenolic compound, however, flavonoids, volatile oils, distemper terpenes, lignans, glycosides, sapogenins, and other waxes have also been found. The flowers are small and intense yellow, formed by five carpels. The leaves show a resin of intense bitter smell. Phenols are mostly found in

**FIGURE 7.3**  *Larrea tridentata.*

stems, followed by leaves and at least proportion in roots (Figure 7.4) [66, 70–72]. NGDA has been found to have fungicidal action against phyto-pathogenic fungi such as *Rhizoctonia solani* [73], *Fusarium oxysporum, Puccinia cacabata* and *Pythium* spp. In addition, extracts of this plant have been tested at different concentrations and obtained with different solvents (methanol, ethanol, chloroform) and have been evaluated against potato plagues such as *Acanthoscelides obtectus, Coleoptera: Bruchidae* (Bean Beetle) and *Prostephanus truncatus, Coleoptera: Bostrichidae* (the grain borer), and nematodes such as *Tylenchus, Ditylenchus,* and *Rabditis* [63].

### 7.2.2.5　AGAVE LECHUGUILLA TORREY (LECHUGUILLA)

The *Agave* genus belongs to the *Agavaceae* family and has more than 400 species [74] that inhabits large semi-arid and arid areas of Mexican Northern and is considered as a non-wood forest product [75]. Its chemical

**FIGURE 7.4**　*Larrea tridentata* (this picture shows in detail leaves and flower).

composition includes secondary metabolites of various kinds, such as sterols, steroids ((25R)-spray-5-ene-2 alpha, 3 beta-diol (yuccagenina) and (25R)-5 beta-spirostan-3 beta), flavonoids, homoisoflavonoids, phenolic acids, tannins, volatile coumarins, long chain alkanes, fatty acids and alcohols [74], however, sapogenins (leaves and roots) and steroid saponins are the phytochemicals that distinguish them. Saponins of which are distinguished three main ones: smillagenin, hecogenina, and manogenin [63]. Presence of these saponins confer significant antifungal activity, the glycoalkaloid saponins being responsible for interacting with the ergosterol of the fungal membrane, which is triggered in the formation of aqueous pores that permeate the membrane [76]. Aqueous extracts obtained from *Agave lechugilla* bulb have shown activity against *Fusarium oxysporum* (Figure 7.5) [63]. Its use is mainly in production and textile applications. Its fiber is used to make cleaning brushes, furniture, carpets, and also with mixtures of thermoplastic resins is employed for building material [75]. It

**FIGURE 7.5**  *Agave lechuguilla.*

is widely used for fibers elaboration (blends with polyesters) used in silos and granaries and as fodder for rabbits [63] but they are also of interest in the food industry for its potential as nutraceutical, prebiotic activity and natural sweetener, although its use in these aspects dates back to ancient times used as a fermented drink (honey water) and as a source of fiber [74], and recently because of its rich cell content, lechuguilla has been proposed as a low-cost biological matrix for the removal of heavy metals in water, and saponins serve as chelants that sequester heavy metals such as Pb, Cr, Cu, Cd, and Zn [77].

### 7.2.2.6    OPUNTIA FICUS-INDICA

Dark green plant covered with a layer. Its height is estimated between 3 to 5 m with stalks 60–150 cm wide and 30–60 cm long. The spines are whitish, short, and weak, although they are regularly absent or are shown in small numbers by each areola (Figure 7.6) (*Opuntia ficus-indica*). The flora is orange to yellow. Its fruit is edible, sweet, and juicy (Figure 7.7). The metabolites that had been reported in this plant species are: gallic acid, terpenes, and saponins [78].

### 7.2.2.7    CAPSICUM SPP. (PEPPER)

Capsicum belongs to the *Solanaceae* family (Mexico has about 40 species of *Capsicum* which are mostly small shrubs [79]. The chemical compound that distinguishes them is capsaicin [80] (N-vanillyl-8-methyl–6-nonen-amide), the main component of capsaicinoids. It is an alkaloid present in different species of the *Capsicum* genus and this compound is responsible for the pungency and pharmacological and physiological properties. Other secondary metabolites such as flavonoids, polyphenols [81], steroidal alka-loids, some glycosides, phenolic acids, vitamin A, carotenoids, tocoph-erols, and ascorbic acid has been reported [82]. Its seeds contain sterols, fatty acids, triterpenes, furostanol saponin derivatives (responsible for antimicrobial activity), and organic acids [83]. This plant has been tested as antimicrobial source against *Escherichia coli, Pseudomonas sola-nacearum, Bacillus subtilis, Clostridium sporogenes, Clostridium tetani, Streptococcus pyogenes, Listeria monocytogenes, Staphylococcus aureus,*

**FIGURE 7.6**   *Opuntia ficus.*

**FIGURE 7.7**   Parts of *Opuntia ficus* fruit.

and *Salmonella enterica Typhimurium,* using the characteristic compound capsaicin with other compounds that acts synergistically with capsaicin [63, 70].

## 7.2.2.8 YUCCA SPP. (DESERT'S PALM)

The main phytochemicals in this plant are steroidal sapogenins, flavonoids, terpenoids, carbohydrates, fats, tannins, and hydrocarbons. This plant is used in the cosmetic industry as emollient and for its antimicrobial effects (Figure 7.8). In the mining industry, it is used to separate gold. On the other hand, it also serves to pre-formulate emulsions used in development of photographic films. In biotechnology is used for bioremediation of contaminated soils and as an adjuvant for nitrogen retention (35%), inhibiting urease enzyme. In addition, this plant has shown in vitro inhibition of cercariae (larvae) growth [63]. Most of the plants have been used as extracts, but use of pure phytochemicals is also mentioned.

**FIGURE 7.8**    (a) *Yucca filifera.* (b) *Yucca carnerosana.*

## 7.3  BIOACTIVE MOLECULES

These molecules have been developed over thousands of years of evolution to promote defenses to antagonist organisms such as viruses, bacteria, and fungi. These compounds are widely distributed in fruits, vegetables, legumes, whole grains, nuts, seeds, fungi, herbs, and spices and in vegetable drinks such as wine and tea [84].

Phytochemicals are divided into two categories: [85] -Primary Metabolites which are essential substances for cellular development (carbohydrates, lipids, proteins, amino acids, nucleic acids) and secondary metabolites, substances that increase organism capacity of survival, granting qualities as protection to interact with the environment. Some examples are phenolic acids, terpenes, alkaloids, etc. They are defined as bioactive substances that confer pharmacological/toxicological effects on humans and animals [54, 86].

### 7.3.1  PHENOLIC COMPOUNDS

In plants, phenolic compounds act directly on sensory properties such as taste and odor, and also provide resistance against predators [87]. In addition, these compounds are related on germination, growth, and reproduction processes, acting as regulators, although their main characteristic is prevention against damage caused by oxidative stress. These compounds represent approximately one-third of the human diet [85]. The metabolic routes for its formation are shikimate, pentose phosphate and phenylpropanoide pathways [40]. These compounds have many properties such as: antiallergics, antiinflammatories, antioxidants, antithrombotics, cardioprotectors, and vasodilators [88, 89] The potential of phenolic compounds is directly correlated with circumstances and elements such as: origin of the vegetal source, culture conditions, vegetal harvest time, type of extraction and solvent used during extraction [40]. These secondary metabolites distributed throughout the plant kingdom are described as molecules with at least one aromatic ring substituted by one or more hydroxyl groups. They may also contain (or not) methoxy and glycosyl groups [90].

## 7.3.1.1   PHENOLIC ACIDS

Phenolic acids are also an important class of phenolic compounds that have bioactive functions and are found in plants and foods [91]. They can be bound by ester, ether or acetal bonds to different plant components or present also in their free form. Chemically, phenolic acids are composed by a benzene ring, a carboxyl group and one or more substitutions of hydroxyl and/or methoxyl groups and their structures may be a single phenolic molecule including a complex polymer with high molecular mass [40]. They are usually grouped into two categories according to their structure [92].

- **Hydroxybenzoic acid:** contain seven carbon atoms. Examples of this type of phenolic acids are p-hydroxybenzoic acid, hierarchic acid, vanillic acid, and gallic acid, presenting a C6-C1 family structure [85].

- **Hydroxycinnamic acid:** contains nine carbon atoms [85]. Examples of this type of phenolic acids are caffeic acid, ferulic acid, hydroxycinnamic acids, synaptic acid, and p-coumaric acid [91].

## 7.3.1.2   FLAVONOIDS

In plants, these compounds act as growth regulators, UV-light protectors, against herbivores, phytopathogens, and protection in cell lesions and form the largest group of phenolic compounds [93, 94]. Flavonoids are structurally described as a skeleton of fifteen carbon atoms with a C6-C3-C6 (Figure 7.9). The structure is composed of two aromatic rings joined by 3 carbon atoms (heterocyclic pyran ring) [95]. One of the major aromatic rings is derived from the metabolic pathway acetate-malonate and the other major ring of the phenylalanine route (shikimate pathway). The substituent chemical groups in the major rings are responsible for differentiation in various classes of flavonoids, such groups can be oxygen, alkyl, glycosylations, acylations or sulfonations [87]. Flavonoids and their classes: flavones, flavonols, flavanones, isoflavones, and anthocyanins have been reported for their different antioxidant, antibacterial, antiviral, and anticancer activities in the human body [87, 95–99]. The chalcones, although with similar structure (aromatic ketones) are not true flavonoids [93], however, it is reported that these molecules have beneficial activities for health, such as anticancer and for inhibition of exoenzymes responsible for fungal invasion mechanisms [96, 100].

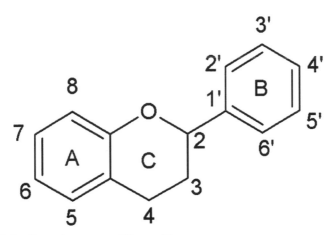

**FIGURE 7.9**   General structure of flavonoid.

## 7.3.1.2.1   *Flavones*

In positions 2 and 3, they have a double bond. At the 4-position of ring C, they have a ketone. At the 5-position of ring A, is a hydroxyl group. At position 7 of ring A, there may be variation (Figure 7.10). Methylation and acylation occur in the B-ring hydroxyl groups, and glycosylation occurs in positions 5 and 7 [101].

## 7.3.1.2.2   *Flavonols*

Of the major classes of flavonoids, they are found in apples, onions, red wine and bananas. The diversity of this class is given by the position of the hydroxyl group (position 3 of ring C has a hydroxyl group) and its state is more oxidized than other flavonoids. It can be glycosylated depending on the state of methylation and hydroxylation (Figure 7.10). These variations can be related with specific activities in the human body [100].

### 7.3.1.2.3   Flavanones

These components are distinguished by a saturated chain of three carbons and one oxygen (C4). Glycosylated (disaccharide) is usually found in C7 chemically considered weak acids (Figure 7.10). Flavanones are abundant in citrus fruits (naringenin in grapefruit, hesperetin in oranges and eridicidol in lemons), tomatoes, and peppermint plant [102].

### 7.3.1.2.4   Flavanols

Chemically, they are described as products of the reduction of dihydroflavonols. They have a hydroxyl group in the 3-position and the saturated carbon ring in its entirety (Figure 7.10). They are commonly called catechins [100, 103].

### 7.3.1.2.5   Isoflavones

These compounds are similar to estrogens with hydroxyl groups at carbon four and seven positions (Figure 7.10). They may be found in aglycon form and in the form of acetyl-, or malonyl-, b-glucosides. Its physiological effects are attributed to its resemblance to b-estradiols [101].

### 7.3.1.2.6   Anthocyanins

Chemically, they are aromatic rings attached to a sugar moiety, generally monosaccharides: glucose, galactose, arabinose, and rhamnose and mono-saccharose trisaccharides (Figure 7.10). Depending on the pH, they are revealed as red, purple or blue dye, and in their isolated form they are very unstable and prone to degradation [104]. They are found in stems, flowers, roots, and fruits.

### 7.3.1.2.7   Anthocyanidins (also Aglycones)

Molecules of three aromatic rings, in one of them a substitution of oxygen. The most frequent anthocyanidins in plants are: delphinidin, petunidine, pelargonidin, malvidin, cyanidin, and peonidin. These molecules are

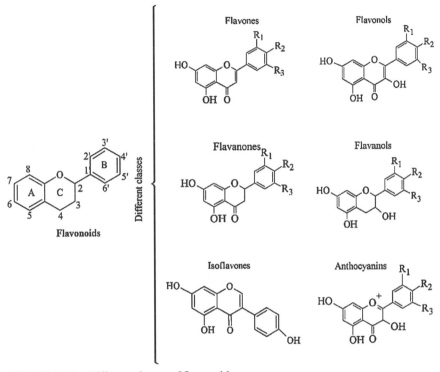

**FIGURE 7.10** Different classes of flavonoids.

responsible for the red pigmentation of fruits. In general, flavonoids in food industries are added to maximize color and taste, to preserve vitamins and enzymes and to prevent fats oxidation. They replace synthetic dyes in foods and pharmaceuticals because of their bright attractive tone and water solubility [105].

### 7.3.1.2.8 Chalcones

They serve as intermediates in the synthesis of flavonoids. They are described as aromatic ketones formed by a pair of phenyl rings joined by a 3-carbon structure [100] (Figure 7.11).

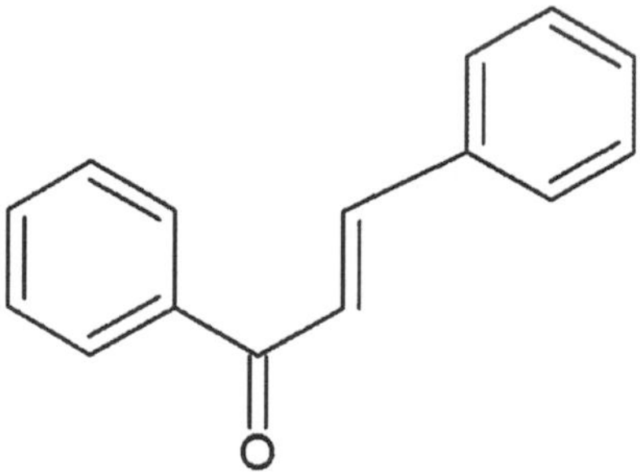

**FIGURE 7.11**　Structure basic chalcone.

## 7.3.1.3　TANNINS

The most important effects of tannins on the human organism are its metal ion-sequencing chelator potential, its ability to precipitate proteins and its antioxidant potential. Because of the wide range of biological activity of these compounds, the food industry has incorporated them into products such as fish, pasta, ice cream, meat, and some milk derivatives, since its antimicrobial activity resides on its capacity to complex with proteins and to inhibit some enzymes and thus it gives resistance to the microbial attack [106].

Proanthocyanidins (condensed tannins) are polymeric flavonoids [85], which are the third major category within the group of polyphenolic compounds and are molecules with high molecular weight that have wide structural variation. Tannins are classified into two groups:

- **Hydrolyzable tannins:** contain a glucose or polyhydroxylated alcohol center partially or totally esterified with gallic acid or hexa-hydroxydiphenic acid; and

- **Non-hydrolyzable tannins:** are polymeric repeats of catechin [85].

### 7.3.1.3.1　Condensed Tannins

The condensed tannins are a group of oligomers and polyhydroxyan-avan–3-ol polymers linked by carbon-carbon bonds (Figure 7.12.). The interaction of proanthocyanidins (condensed tannins) has important nutritional and physiological effects, such as the complexing of proteins and other molecules. This reactivity is given by the substituent groups R, R2, and R3. Substitutions of a hydroxyl group at the 5-position (ring B) increase the ability to complex proteins [107].

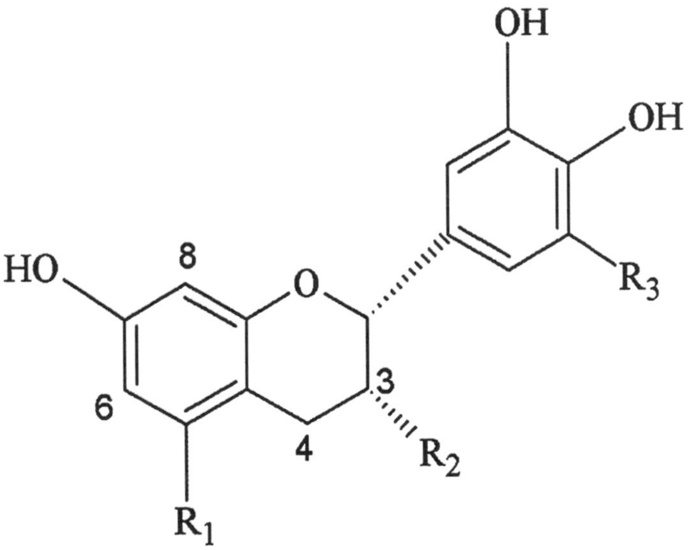

**FIGURE 7.12** Unit basic form condensed tannins.

### 7.3.1.3.2 Hydrolyzable Tannins

Hydrolyzable tannins are secondary metabolites that are water-soluble and have molecular masses of 500–5000 Daltons. Chemically, they are esters of Gallic acid and glucose (Figure 7.13.). The property that distinguishes them and which is of importance in the food industry is to give specific characteristics to foods, mainly astringency, and also for their antibacterial potential. This sensation in food is a result of the complexation of saline proteins rich in proline and hydrolyzable tannins. In the pharmacological field, they are interesting for their antitumor, anti-mutagenic, anti-diabetic, anti-proliferative, and anti-mycotic properties [108].

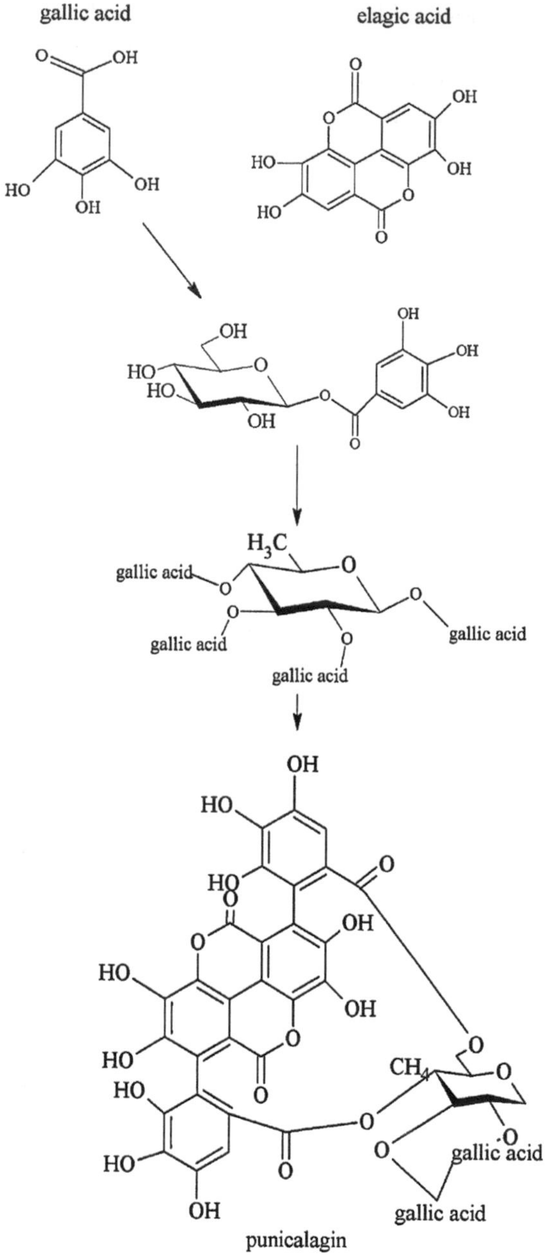

**FIGURE 7.13** Hydrolyzable tannins.

## 7.3.2 STEROLS

Sterols are a fundamental part of the membranes in eukaryotic cells, in addition to helping other tasks such as growth. These compounds occur in different forms: cholesterol is the predominant sterol in mammals, ergosterol in yeast and other fungi, and phytosterols are the main sterols in plants [109]. Plant sterols resemble cholesterol in mammals, however, plants produce complex sterols that differ primarily in the side chain, the number and position of double bonds in the chain or ring [110, 111] ring system cyclopentane-per-hydrophenanthrene with a hydroxyl group (C3) and a side chain (length of between 8–10 carbons) which is attached to C17 (Figure 7.14). The main sterols of the plants are β-sitosterol, stigmasterol, and campesterol (Figure 7.15) [109, 112]. The sources of phytosterols are vegetable oils, pastes, breads, vegetables, and margarines. Approximately, 300 mg of phytasterols are consumed in the common diet (Western culture) [113].

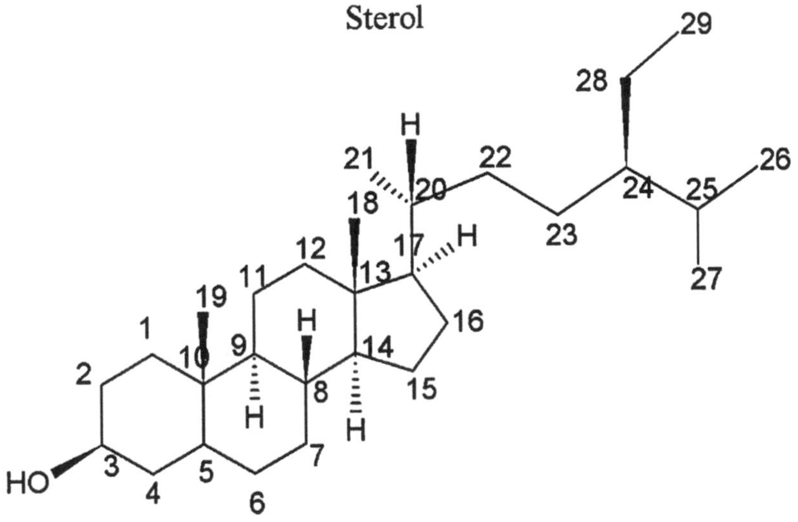

**FIGURE 7.14**  Structure basic sterol.

In the food industry, sterols of plant origin are used for the progress of functional foods. Upon incorporation (sterols and stanols) undergo esterification to increase the solubility in the matrix in which it is introduced

[85]. Phytosterols are studied for their ability to lower plasma cholesterol levels and the FDA has approved this health claim for phytosterols such as a cholesterol-lowering agent [111, 114].

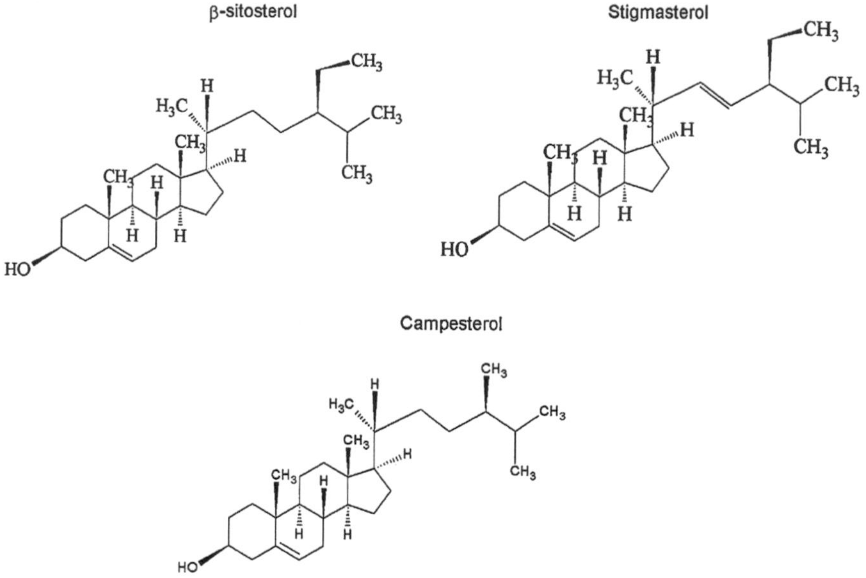

β-sitosterol

Stigmasterol

Campesterol

**FIGURE 7.15** The main sterols of the plants.

## 7.3.3 ESSENTIAL OILS

These compounds are found in flowers, seeds, buds, leaves, fruits, roots or barks and obtained by distillation [115]. Chemically, they are mixture of different compounds of low molecular weight, such as terpenes (mono-, sesqui-, and diterpenes), ketones, phenols, alcohols, acids, epoxides, amines, sulphides, esters, and aldehydes. These compounds are considered as GRAS substances, they are widely used mainly as aids in flavoring. Its use in food, confectionery, and beverages is mainly to increase their flavors [17, 116]. In addition, essential oils are used to reduce food borne pathogens, [117,118] infectious agents such as *Salmonella spp., Escherichia coli* O157: H7, *Campylobacter,* and *Listeria monocytogenes* [39,

119]. Although, it is worth mentioning that antimicrobial use is limited by taste appreciations [7]. The antimicrobial potential and the different mechanisms of action of essential oils are attributed to the chemical structure of its different components (monoterpenes, sesquiterpenes, alcohols, aldehydes, esters, ethers, ketones, phenols, oxides, phenylpropenes, and specific substances containing sulfur or nitrogen), the interaction among them (synergic action) and the concentration of each component [10]. Some researchers propose that the mechanism by which the essential oils are obtained affects their antimicrobial capacity because the lipophilic properties that give them the ability to penetrate through the bacterial membranes and inhibit vital functions of the cell. On the other hand, phenolic nature can also alter the permeability of microbial cells [17].

## 7.3.4  LIGNANS

In the IUPAC canons, lignans are described as dimers connected to coniferyl alcohols [120]. Primarily the term was coined in 1936 by Haworth to designate a group of phenylpropanoid dimers which at C6-C3 bind to a central carbon of their propyl side chains [121]. The classification is based on the way in which the oxygen atom joins the skeleton in addition to the cyclization of the molecule, being these compounds classified into eight subgroups: furofuro, furan, dibenzylbutane, dibenzylbutyrolactone, aryltetralin, arylnaphthalene, dibenzocyclooctadiene, and dibenzylbutyrolactol [120, 122]. They are synthesized from phenylpropanoid monomers (oxidative dimerization) and a coniferyl alcohol by the assistance of proteins with reductase-dehydrogenase-hydroxylase-methyltransferase-oxygenase activity [123].

Lignans are abundant in the plant kingdom, forming part of rhizomes, stems, roots, leaves, seeds, and fruits. They are secondary metabolites of plants, the role in vegetables is to provide defense against various biological pathogens, although each of these functions are not yet very clear [124]. The biological activities that highlight them are antitumor activity, antiviral activity, inhibition of the activity of enzymes such as cytochrome oxidase and cyclic adenosine monophosphate phosphodiesterase, cardiovascular effects, allergenicity, and toxic activities to insects, microbes, fungi, fish, plants, and mammals [123].

## 7.3.5   NATURAL PIGMENT

The recent high demand of customers for natural dyes and food industry concerns about safety and health benefits that these compounds may give to food, make natural pigment an interesting research topic. In addition, these compounds may improve appearance [125].

## 7.3.6   SAPONINS

These compounds are a large group of glycosides formed by an aglycone unit (hydrophobic, sapogenin) and sugar (hydrophilic, glucone) [74] They are classified into two groups according to the nature of the aglycone skeleton: steroid saponins (a spirostane skeleton C27, generally a structure of aglycone) and glycogen (glycoprotein, glycine). Six rings and triterpenoid saponins (a C30 skeleton comprising a pentacyclic structure) [126]. In water, colloidal solutions and foam (when shaken) are formed as a result of the combination of polar and non-polar structural elements in their molecules [127]. Its use in pharmacy is to serve as a molecule of semi-synthesis of steroidal drugs. The outstanding biological properties are anti-inflammatory, antifungal, antimicrobial, antiparasitic, and antiviral activities [126]. In the food industry, saponins have found wide applications in beverages and confectionery [127].

## 7.4   HOW DO BIOACTIVE COMPOUNDS WORK?

It is worth mentioning that in spite of the known antibacterial capacity (even of its antioxidant power) of diverse molecules obtained from natural sources, the mechanism of precise action on its operation remains unknown or described in detail, several mechanisms of action are proposed, but are still not well defined [34]. However, different ways for their biological function have been proposed (Figure 7.16) [128]. While, for the antimicrobial activity, some mechanisms of action are proposed such as: OH groups that form hydrogen bonds with effect that modifies the enzymes, changing their intracellular functions, interaction with membrane enzymes, changing the cellular form, which causes decrease or loss of membrane integrity and modification of membrane permeability and rupture of the altered cytoplasm (formation of granules) [10]. Polyphenols have effect on bacteria membranes absorption, altering its functionality and inducing flow of potassium ions, causing alteration, and loss of cytoplasmic content. Besides, it is reported formation of hydroxyl peroxides by presence of polyphenols [7].

Catechins are also active on the outer polar part of the lipid bilayer in liposomes and cause membrane disruption. Similarly, terpenes cause loss in membrane integrity, and dissipating proton strength and altering lipid structures [7]. In some plants (for example coffee), the chelating properties of the metal are proposed as antimicrobial activity, being the responsible for its bactericidal activity [34]. On the other hand, the suitability of hydrophobic nature compounds to dissolve in the lipid portion of cytoplasmic membranes, is the reason for their activity, triggering alterations in permeability, osmotic intracellular imbalance, malfunction of organelles, leakage of vital molecules such as adenosine triphosphate (ATP) and inevitably cell death [10].

Regarding the antioxidant power, some polyphenol compounds act as stimulating expression of genes involved in the production of antioxidant

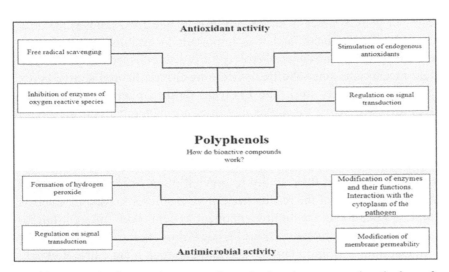

**FIGURE 7.16** The diagram shows some forms that have been proposed on the form of action of the polyphenols [7, 9, 10, 85, 95, 123, 128]. The mechanisms shown, address only two biological activities exhibiting this class of compounds: antioxidant activity and antimicrobial activity.

enzymes. Another pathway is through regulation of expression of genes involved on cyto-protection process [85]. With regard to essential oils, its mechanism of action is associated with the ability to interact with the pathogen cytoplasm. Its mode of action seems to be closely related to the solubility of the compound, and the antifungal activity of essential oils is associated with the phenol content of compounds such as monoterpene [9].

### 7.4.1   PLANT BIOACTIVE COMPOUNDS IN FOOD INDUSTRIES: BIO ACCESSIBILITY AND BIOAVAILABILITY OF FUNCTIONAL INGREDIENTS IN FOODS

It is well documented that several natural compounds present in plants exhibit antimicrobial activities and serve as antimicrobial agents against foodborne pathogenic microorganisms [39, 129]. Therefore, special interest has been directed to extracts of vegetal origin as a starting point to find new compounds with antioxidant and antimicrobial capacity for use in food [36]. Bioactive extracts or purified compounds from plant sources can be added to foods, obtaining new functional products and thus open new commercial fields [85]. The availability of bioactive compounds present in plant sources is an opportunity for the food industry to offer new and safer products, and it is important to establish that although there are "natural" substances involved, it is also indispensable to include safety precautions since the fact of coming from a natural source is not a guarantee of safety [40]. The FDA has defined bioavailability as the extent to which active substrates in a drug are integrated into the site of action. The administration of bioactive compounds requires formulations and food techniques to protect them and maintain their active molecular form obtaining to the moment of consumption, thus ensuring their physiological benefits and their ability to prevent diseases [85]. On the other hand, extrapolation of the results from *in vitro* studies to food products is complex, since foods are multi-component systems with different micro-environments and interconnections [7, 93, 97].

    In addition, it is important to note that although plant extracts clearly have great potential as antimicrobial substances, variations in the quality and quantity of the phytochemical profile of extracts are a significant obstacle to their use, showing a variable efficacy (lack of reproducibility) depending on the models chosen, laboratory and other factors (extraction methods) [7]. It should be noted that different international organizations (guidelines of the Antimicrobial Susceptibility Testing Committee, and National Institutes of Health (NIH)) have been able to propose effective guidelines to standardize *in vitro* antibacterial evaluation methods. However, the antibacterial evaluation of plant extracts is not clear [49]. For another side, it is also necessary to point out the importance of regulation of antioxidant compounds in foods to avoid some adverse effects of polyphenols such as: formation of strong complexes with dietary, salivary, and digestive proteins. A specific example of auto-oxidation that is reflected in loss of color in processed vegetables or excessive changes of food texture [40]. It is known that to date, there is no single standard method (or considered as such) for extraction of bioactive compounds of vegetal origin [49].

Amount and quality of bioactive molecules that are extracted from plant sources depend on selection of extraction method [88]. There is a wide variety of extraction methods and solvents, from conventional techniques (soxhlet extraction, maceration, and hydro-distillation) to the most innovative ones and considered green (ultrasonic assisted extraction) [52, 130] which recently has incremented its popularity over traditional methods ones because of its environmentally friendly nature [131]. In the same way, the extraction of phyto-compounds, (polyphenols and essential oils) assisted by microwave, since it maintains of equal is friendly with the environment and in addition it offers saving of time and energy, applicability to samples smaller and temperatures lower than guarantee a real and effective recovery of phytocomposites [132]. The solvents used may be diethyl ether, ethanol, methanol, dichloromethane, chloroform, acetone, water, and other hydro-alcoholic solvents [49]. Selection of extraction method is mostly related to bioactive compounds classification [85]. For example, hydrophilic or polar (phenolic acids, flavonoids, organic acids, and sugars), lipophilic or non-polar (carotenoids, terpenoids, alkaloids, tocopherols, steroids, and fatty acids).

The main problem for methods standardization to obtaining extracts or powders is the diverse chemical composition of plant extracts which depends on geographical origin of source, plant variety and part of the plant used, plant age and maturation status [90]. Regarding antioxidants, despite current research, little is known about bioavailability [93, 97], *in vivo* studies have been performed in eukaryotic organisms such as fish, rabbits, but few in humans. These studies may clarify whether some are effective after being metabolized or effective in their active form, if there is positive or negative interference when ingested with other nutrients or if there is a condition of the individual that consumes them and even between the doses ingested. This may explain why some plant species, even isolated compounds, are tested without evidence of the true action of their antioxidant potential when tested in a living system [89]. Also, plant extracts are composed of a large variety of chemical compounds, making it difficult to establish a relationship between a compound and a single target in the cell [23].

## 7.5  FUTURE TRENDS

As a source of many characteristics that benefit human health, the search for bioactive compounds in various matrices (fruits, plants, agricultural residues, and agroindustrial waste) [83, 133] has intensified considerably. The use of agroindustrial residues for biological transformation processes provides a wide range of alternative substrates [133–135], which in addition to serving as new extraction matrices for phytocompounds, also help to combat pollution problems [137, 138]. With respect to the above, the use of processes to obtain phytochemicals in bulk from agroindustrial residues as a raw material has spread throughout the world [139]. One of the innovators and alternatives to ensure food safety, besides the use of natural antimicrobials such as plant extracts [106, 140, 141], essential oils, and organic acids, there is also mentioned the use of bacteriocins. In addition, the trend of using natural products rather than synthetic products is on the rise, not only in the food area, but also in other industries such as cosmetics and pharmaceuticals [41, 46, 134, 136].

Byproducts produced after different processes in the food industry are important targets for the opportunity to extract phenolic compounds. For example, studies indicate that shells and seeds produced after the processing of various fruit juices (representing 50% of the total weight) have higher polyphenolic content than even the edible parts. A similar case is observed in grape skins, where the wine is produced. In the case of apple pomace, the husks have 3300 mg (100g) in phenolic compounds [104]. Currently, with high levels of production, waste disposal is often a problem, despite its traditional use as biofertilizer or as feed, demands, and yields can be oscillating, adding aggravating legal restrictions.

Therefore, reuse of these byproducts for obtaining bioactive compounds is a cheap, efficient, and extremely friendly and environmentally friendly option [142].

## 7.6 CONCLUSIONS

Interest in plant-based antimicrobials has intensified in recent decades, since further prolonging shelf life and providing food safety to meet the interests and demands of current consumers who prefer food consumption without artificial preservatives. In addition, to covering a nutritional and sensory role, they also provide potential benefits by controlling and preventing diseases of global incidence such as obesity, cardiovascular disorders, cancer, and neurodegenerative diseases. On the other hand, appearance of bacteria resistant to various drugs makes imperative the need to find new antibacterial agents, and phytochemicals are the ideal option for development of new drugs. The effectiveness of plant extracts to combat or inhibit pathogenic microorganisms is demonstrable, however, in this context, it is important to note that use of these biomolecules represents a challenge and an area of opportunity for the food industry, since the effective administration for its use requires standardized formulations, plus optimization conditions and more effective extraction methods, which ensure the integral functionality of such compounds at the time be arranged in a food matrix. Therefore, future use of these antimicrobial molecules should address aspects such as a detailed, sequential and quantitative studies to evaluate the antimicrobial efficiency of extracts; a study on the intrinsic and extrinsic storage parameters of the product where the antimicrobials could be used; use of techniques and methodologies that protect molecular structures up to the moment of their consumption and action, and finally an appropriate regulation for the use of those compounds, as well as, the study and elucidation of specific molecular mechanisms of action of these photocomposes in the target cells. Based on what has been exposed in this chapter, it is evident that use of agroindustrial residues as well as undervalued crops in the semi-arid regions as matrices for obtaining antimicrobials, makes the search even more valuable and interesting since these compounds are friendly with the environment and have low economic cost.

## ACKNOWLEDGMENTS

Rosario Estrada Mendoza thanks the National Council of Science and Technology of Mexico (CONACyT) for the financial support provided during her MSc studies under the scholarship agreement number 611520. Financial support was received from SAGARPA-CONACyT through the project: "Obtention, purification and scale-up of bioactive compounds extraction with industrial value, obtained using advance technology of extraction from cultures, byproducts and little valued natural resources" SAGARPA-CONACyT 2015-4-266936.

## CONFLICT OF INTEREST

Authors express no conflict of interest.

## KEYWORDS

- **foodborne diseases**
- **microbial resistance to antibiotics**
- **pathogen and spoiler microorganisms**
- **polyphenols**

## REFERENCES

1. Hu, K., Renly, S., Edlund, S., Davis, M., & Kaufman, J., (2016). A modeling framework to accelerate foodborne outbreak investigations. *Food Control.*, *59*, 53–58.
2. Fleckenstein, J. M., Bartels, S. R., Drevets, P. D., Bronze, M. S., & Drevets, D. A., (2010). Infectious agents of food- and water-borne illnesses. *Am. J. Med. Sci.*, *340*(3), 238–246.
3. Santhi, L. P., Sunkoji, S., Siddiram, S., & Sanghai, S. S., (2012). Patent research in salmonellosis prevention. *Int. Food Res. J.*, *45*(2), 809–818.
4. Jarriyawattanachaikul, W., Chaveerach, P., & Chokesajjawatee, N., (2016). Antimicrobial activity of Thai-herbal plants against foodborne pathogens *E. coli, S. aureus,* and *C. jejuni. Agric. Agric. Sci. Procedia.*, *11*, 20–24.
5. Linscott, A. J., (2011). Foodborne illnesses. *Clin. Microbiol. Newsl.*, *33*(6), 41–45.

6. Velusamy, V., Arshak, K., Korostynska, O., Oliwa, K., & Adley, C., (2010). An overview of foodborne pathogen detection: In the perspective of biosensors. *Biotechnol. Adv., 28*(2), 232–254.
7. Negi, P. S., (2012). Plant extracts for the control of bacterial growth: Efficacy, stability and safety issues for food application. *Int. J. Food Microbiol., 156*(1), 7–17.
8. Srey, S., Jahid, I. K., & Ha, S. Do., (2013). Biofilm formation in food industries: A food safety concern. *Food Control., 31*(2), 572–585.
9. De Ramos-García, M. D. L., Bautista-Baños, S., & Barrera-Necha, L. L., (2010). Antimicrobial Compounds Added in Edible Coatings for Use in Horticultural Products. *Rev. Mex. Fitopatol., 28*(1), 44–57.
10. Da Cruz Cabral, L., Fernández, P. V., & Patriarca, A., (2013). Application of plant derived compounds to control fungal spoilage and mycotoxin production in foods. *Int. J. Food Microbiol., 166*(1), 1–14.
11. Yen, P. K., (2003). Preventing harm from foodborne illness. *Geriatric Nursing, 24*(6), 376–377.
12. Cespedes, C. L., Alarcon, J., Aqueveque, P. M., Lobo, T., Becerra, J., Balbontin, C., et al., (2015). New environmental friendly antimicrobials and biocides from Andean and Mexican biodiversity. *J. Environ. Res., 142*, 549–562.
13. Tauxe, R. V., Doyle, M. P., Kuchenmüller, T., Schlundt, J., & Stein, C. E., (2010). Evolving public health approaches to the global challenge of foodborne infections. *Int. J. Food Microbiol., 139*(1), 16–28.
14. Greppi, A., & Rantsiou, K., (2016). Methodological advancements in foodborne pathogen determination: From presence to behavior. *Current Opinion in Food Science, 8*, 80–88.
15. De Koster, C. G., & Brul, S., (2016). MALDI-TOF MS identification and tracking of food spoilers and foodborne pathogens. *Curr. Opin. Food Sci., 10*, 76–84.
16. Remenant, B., Jaffres, E., Dousset, X., Pilet, M. F., Zagorec, M. 2015. Bacterial spoilers of food: behavior, fitness and functional properties. *Food Microbiol., 45*(PA), 45–53.
17. Calo, J. R., Crandall, P. G., O'Bryan, C. A., & Ricke, S. C., (2015). Essential oils as antimicrobials in food systems – A review. *Food Control., 54*, 111–119.
18. Azeredo, D. R. P., Alvarenga, V., Sant'Ana, A. S., & Sabaa Srur, A. U. O., (2016). An overview of microorganisms and factors contributing for the microbial stability of carbonated soft drinks. *Food Res. Int., 82*, 136–144.
19. Barbieri, R., Coppo, E., Marchese, A., Daglia, M., Sánchez, E., Nabavi, S. F., et al., (2017). Phytochemicals for human disease: An update on plant-derived compounds antibacterial activity. *Microbiol. Res., 196*, 44–68.
20. Mann, E., Wetzels, S. U., Pinior, B., Metzler-Zebeli, B. U., Wagner, M., & Schmitz-Esser, S., (2016). Psychrophile spoilers dominate the bacterial microbiome in musculature samples of slaughter pigs. *Meat Sci., 117*, 36–40.
21. Sade, E., Penttinen, K., Bjorkroth, J., & Hultman, J., (2017). Exploring lot-to-lot variation in spoilage bacterial communities on commercial modified atmosphere packaged beef. *Food Microbiol., 62*, 147–152.

22. Maifreni, M., Frigo, F., Bartolomeoli, I., Buiatti, S., Picon, S., & Marino, M., (2015). Bacterial biofilm as a possible source of contamination in the microbrewery environment. *Food Control., 50*, 809–814.

23. Haagsma, J. A., Polinder, S., Stein, C. E., & Havelaar, A. H., (2013). Systematic review of foodborne burden of disease studies: Quality assessment of data and methodology. *Int. J. Food. Microbiol., 166*(1), 34–47.

24. Rocourt, J., (2014). Public health measures: Foodborne disease outbreak investigation. *Encyclopedia of Food Safety.* Elsevier. Paris, France, *4*, 89–97.

25. Backert, S., Tegtmeyer, N., Cróinín, T., Boehm, M., & Heimesaat, M. M., (2017). Chapter 1 Human campylobacteriosis. In: Levy, N., & Klein, G., (eds.), *Campylobacter: Features, Detection, and Prevention of Foodborne Disease* (pp. 1–25). Elsevier; Hannover, Germany.

26. Von Altrock, A., Hamedy, A., Merle, R., & Waldmann, K. H., (2013). *Campylobacter spp.* – Prevalence on pig livers and antimicrobial susceptibility. *Prev. Vet. Med., 109*(1–2), 152–157.

27. Castillo, S., Heredia, N., Arechiga-Carvajal, E., & García, S., (2014). Citrus extracts as inhibitors of quorum sensing, biofilm formation and motility of *Campylobacter jejuni. Food Biotechnol., 28*(2), 106–122.

28. Vaquero, M. J. R., Alberto, M. R., & Nadra, M. C. M., (2007). Influence of phenolic compounds from wines on the growth of *Listeria monocytogenes. Food Control., 18*, 587–593.

29. Buchanan, R. L., Gorris, L. G. M., Hayman, M. M., Jackson, T. C., & Whiting, R. C., (2017). A review of *Listeria monocytogenes*: An update on outbreaks, virulence, dose-response, ecology, and risk assessments. *Food Control., 75*, 1–13.

30. Caprioli, A., Morabito, S., & Scavia, G., (2014). Bacteria shiga toxin-producing *Escherichia coli* and other pathogenic *Escherichia coli. Encyclopedia of Food Safety.* Simon Holt, Rashmi Phadnis; Elsevier; Rome, Italy, *1*, 417–423.

31. Word Health Organization, (2007–2015), Chapter: 4. Hazard-specific methodology, In: *WHO Estimates of the Global Burden of Foodborne Diseases: Foodborne Disease Burden Epidemiology Reference Group.* Word Health Organization, Switzerland, (2015), 29–60. Word Health Organization.https://www.cabdirect.org/cabdirect/abstract/20163387522, most recent access: 7 October, 2017.

32. Manfreda, G., & De Cesare, A., (2016). Novel food trends and climate changes: Impact on emerging foodborne bacterial pathogens. *Curr. Opin. Food Sci., 8*, 99–103.

33. Alvarez-Ordóñez, A., Broussolle, V., Colin, P., Nguyen-The, C., & Prieto, M., (2015). The adaptive response of bacterial foodborne pathogens in the environment, host and food: Implications for food safety. *Int. J. Food Microbiol., 213*, 99–109.

34. Runti, G., Pacor, S., Colomban, S., Gennaro, R., Navarini, L., & Scocchi, M., (2015). *Arabica coffee* extract shows antibacterial activity against *Staphylococcus epidermidis* and *Enterococcus faecalis* and low toxicity towards a human cell line. *LWT Food Sci. Technol., 62*(1), 108–114.

35. Thielmann, J., Kohnen, S., & Hauser, C., (2017). Antimicrobial activity of *Olea europaea* Linné extracts and their applicability as natural food preservative agents. *Int. J. Food Microbiol., 251*, 48–66.

36. Mahajan, D., Bhat, Z. F., & Kumar, S., (2015). Pomegranate (*Punica granatum*) rind extract as a novel preservative in cheese. *Food Biosci.*, *12*, 47–53.

37. Mendez, M., Rodríguez, R., Ruiz, J., Adame, D., Castillo, F., Castillo, F. D., et al., (2012). Antibacterial activity of plant extracts obtained with alternative organics solvents against foodborne pathogen bacteria. *Ind. Crops Prod.*, *37*(1), 445–450.

38. Chahad, O. B., El Bour, M., Calo-Mata, P., Boudabous, A., & Barros-Velazquez, J., (2012). Discovery of novel biopreservation agents with inhibitory effects on growth of foodborne pathogens and their application to seafood products. *Res. Microbiol.*, *163*(1), 44–54.

39. Kotzekidou, P., Giannakidis, P., & Boulamatsis, A., (2008). Antimicrobial activity of some plant extracts and essential oils against foodborne pathogens in vitro and on the fate of inoculated pathogens in chocolate. *LWT Food Sci. Technol.*, *41*(1), 119–127.

40. Moure, A., Cruz, J. M., Franco, D., Domínguez, J. M., Sineiro, J., Domínguez, H., et al., (2001). Natural antioxidants from residual sources. *Food Chem.*, *72*(2), 145–171.

41. Wong-Paz, J. E., Contreras-Esquivel, J. C., Rodríguez-Herrera, R., Carrillo-Inungaray, M. L., López, L. I., Nevárez-Moorillón, G. V., et al., (2015). Total phenolic content, in vitro antioxidant activity and chemical composition of plant extracts from semiarid Mexican region. *Asian Pac. J. Trop. Med.*, *8*(2), 104–111.

42. Zarai, Z., Boujelbene, E., Ben Salem, N., Gargouri, Y., & Sayari, A., (2013). Antioxidant and antimicrobial activities of various solvent extracts, piperine and piperic acid from *Piper nigrum*. *LWT Food Sci. Technol.*, *50*(2), 634–641.

43. Murray, B. G., (2017). Chapter: Plant diversity, conservation and use. *Encyclopedia of Applied Plant Sciences* (Vol. 1, pp. 289–308). Elsevier. (Second Edition), Aucklan, New Zealand.

44. Prashantkumar, P., & Gm, V., (2008). Traditional knowledge on medicinal plants used for the treatment of skin diseases in Bidar district, Karnataka. *Indian J. Tradit. Knowl.*, *Vol. 7*(2), 273–276.

45. Bhatia, D., Mandal, A., Nevo, E., & Bishayee, A., (2015). Apoptosis-inducing effects of extracts from desert plants in HepG2 human hepatocarcinoma cells. *Asian Pac. J. Trop. Biomed.*, *5*(2), 87–92.

46. Wong P. J. E., Muñiz, M. D. B., Martínez, Á. G. C. G., Belmares, C. R. E., & Aguilar, C. N., (1015). Ultrasound-assisted extraction of polyphenols from native plants in the Mexican desert. *Ultrason. Sonochem.*, *22*, 474–481.

47. Verástegui, M. A., Sánchez, C. A., Heredia, N. L., & García-Alvarado, J. S., (1996). Antimicrobial activity of extracts of three major plants from the Chihuahuan desert. *J. Ethnopharmacol.*, *52*(3), 175–177.

48. Tiwari, B. K., Valdramidis, V. P., O'Donnell, C. P., Muthukumarappan, K., Bourke, P., & Cullen, P. J., (2009). Application of natural antimicrobials for food preservation. *J. Agric. Food Chem.*, *57*(14), 5987–6000.

49. Tapia-Torres, Y., López-Lozano, N. E., Souza, V., García-Oliva, F., (2015). Vegetation-soil system controls soil mechanisms for nitrogen transformations in an oligotrophic Mexican desert. *J. Arid. Environ.*, *114*, 62–69.

50. Ezcurra, E., & Mellink, E., (2013). Desert ecosystems. *Encyclopedia of Biodiversity* (Vol. 2, pp. 457–478). Elsevier. Ensenada, BC, Mexico. (Second Edition).

51. Zolotokrylin, A. N., Titkova, T. B., & Brito-Castillo, L., (2016). Wet and dry patterns associated with ENSO events in the Sonoran Desert from, 2000–2015. *J. Arid. Environ.*, *134*, 21–32.

52. Azmir, J., Zaidul, I. S. M., Rahman, M. M., Sharif, K. M., Mohamed, A., Sahena, F., et al., (2013). Techniques for extraction of bioactive compounds from plant materials: A review. *J. Food Eng.*, *117*(4), 426–436.

53. Sharma, A., Flores-Vallejo, R., Del, C., Cardoso-Taketa, A., & Villarreal, M. L., (2016). Antibacterial activities of medicinal plants used in Mexican traditional medicine. *J. Ethnopharmacol.*, 1–66.

54. Silva, M. P., Leonardo, A., Piazza, L. A., López, D., López-Rivilli, M. J., Turco, M. D., et al., (2012). Phytotoxic activity in *Flourensia campestris* and isolation of (-)-hamanasic acid A as its active principle compounds. *Phytochemistry*, *77*, 140–148.

55. Piazza, L. A., López, D., Silva, M. P., López, R. M. J., Cantero, J. J., Tourn, G. M., & Scopel, A. L., (2014). Characterization of quaternary ammonium compounds in *Flourensia xerophytic* communities and response to UV-B radiation. *S. African J. Bot.*, *94*, 14–23.

56. Rios, M. Y., Estrada-Soto, S., Flores-Morales, V., & Aguilar, M. I., (2013). Chemical constituents from *Flourensia resinosa* S.F. Blake (*Asteraceae*). *Biochem. Syst. Ecol.*, *51*, 240–242.

57. Mata, R., Bye, R., Linares, E., Macías, M., Rivero-Cruz, I., Pérez, O., et al., (2003). Phytotoxic compounds from *Flourensia cernua*. *Phytochemistry*, *64*(1), 285–291.

58. Estell, R. E., Anderson, D. M., & James, D. K., (2016). Defoliation of *Flourensia cernua* (tarbush) with high-density mixed-species stocking. *J. Arid. Environ.*, *130*, 62–67.

59. Jasso de Rodríguez, D., Hernández-Castillo, D., Angulo-Sánchez, J. L., Rodríguez-García, R., Villarreal, Q. J. A., & Lira-Saldivar, R. H., (2007). Antifungal activity *in vitro* of *Flourensia* spp. extracts on *Alternaria* sp., *Rhizoctonia solani*, and *Fusarium oxysporum*. *Ind. Crops Prod.*, *25*(2), 111–116.

60. Jasso De Rodríguez, D., Salas-Méndez, E., De, J., Rodríguez-García, R., Hernández-Castillo, F. D., Días-Jiménez, M. L. V., Sáenz-Galindo, A., et al., (2017). Antifungal activity in vitro of ethanol and aqueous extracts of leaves and branches of *Flourensia* spp. against postharvest fungi. *Ind. Crops Prod.*, 0–1.

61. De León-Zapata, M. A., Pastrana-Castro, L., Rua-Rodríguez, M. L., Alvarez-Perez, O. B., Rodríguez-Herrera, R., & Aguilar, C. N., (2016). Experimental protocol for the recovery and evaluation of bioactive compounds of tarbush against postharvest fruit fungi. *Food Chem.*, *198*, 62–67.

62. De Rodríguez, D. J., García, R. R., Castillo, F. D. H., González, C. N. A., Saenz, A., Quintanilla, J. A. V., et al., (2011). In vitro antifungal activity of extracts of Mexican Chihuahuan desert plants against postharvest fruit fungi. *Ind. Crops Prod.*, *34*(1), 960–966.

63. Rodríguez, D. J. De., Angulo-Sanchez, J. L., & Hernandez-Castillo, F. D., (2006). Chapter 14: An overview of the antimicrobial properties of Mexican medicinal plants. *Adv. Phytomed.*, *3*(C), 325–377.

64. Pletsch, M., & Charlwood, B. V., (1997). Accumulation of diterpenoids in cell and root-organ cultures of *Jatropha* species. *J. Plant Physiol.*, *150*(1–2), 37–45.

65. Araujo-Espino, D. I., Zamora-Perez, A. L., Zuniga-González, G. M., Gutiérrez-Hernández, R., Morales-Velazquez, G., & Lazalde-Ramos, B. P., (2017). Genotoxic and cytotoxic evaluation of *Jatropha dioica* Sessé x Cerv. by the micronucleus test in mouse peripheral blood. *Regul. Toxicol. Pharmacol., 86*, 260–264.

66. Gupta, D., Bleakley, B., & Gupta, R. K., (2007). Dragon's blood: Botany, chemistry and therapeutic uses. *J. Ethnopharmacol., 115*(3), 361–380.

67. Osorio, E., Flores, M., Hernández, D., Ventura, J., Rodríguez, R., & Aguilar, C. N., (2010). Biological efficiency of polyphenolic extracts from pecan nuts shell (*Carya Illinoensis*), pomegranate husk (*Punica granatum*) and creosote bush leaves (*Larrea tridentate* Cov.) against plant pathogenic fungi. *Ind. Crops Prod., 31*(1), 153–157.

68. Bhuyan, D. J., Van Vuong, Q., Chalmers, A. C., Van Altena, I. A., Bowyer, M. C., & Scarlett, C. J., (2015). Microwave-assisted extraction of *Eucalyptus robusta* leaf for the optimal yield of total phenolic compounds. *Ind. Crops Prod., 69*, 290–299.

69. Abou-Gazar, H., Bedir, E., Takamatsu, S., Ferreira, D., & Khan, I. A., (2004). Antioxidant lignans from *Larrea tridentata. Phytochemistry., 65*(17), 2499–2505.

70. Lambert, J. D., Zhao, D., Meyers, R. O., Kuester, R. K., Timmermann, B. N., & Dorr, R. T., (2002). Nordihydroguaiaretic acid: Hepatotoxicity and detoxification in the mouse. *Toxicon., 40*(12), 1701–1708.

71. Lira-Saldívar, R. H., (2003). Current State of Knowledge about Biocidal Properties of Gobernadora [*Larrea tridentata* (D.C.) Coville]. *Rev. Mex. Fitopatol., 21*(2), 214–222.

72. Arteaga, S., Andrade-Cetto, A., & Cárdenas, R., (2005). *Larrea tridentata* (Creosote bush), an abundant plant of Mexican and US-American deserts and its metabolite nordihydroguaiaretic acid. *J. Ethnopharmacol., 98*(3), 231–239.

73. Castillo, F., Hernández, D., Gallegos, G., Mendez, M., Rodríguez, R., Reyes, A., et al., (2010). *In vitro* antifungal activity of plant extracts obtained with alternative organic solvents against *Rhizoctonia solani* Khn. *Ind. Crops Prod., 32*(3), 324–328).

74. Sidana, J., Singh, B., & Sharma, O. P., (2016). Saponins of agave: Chemistry and bioactivity. *Phytochemistry, 130*, 22–46.

75. Pando-Moreno, M., Pulido, R., Castillo, D., Jurado, E., & Jiménez, J., (2008). Estimating fiber for lechuguilla (*Agave lecheguilla* Torr., Agavaceae), a traditional non-timber forest product in Mexico. *For. Ecol. Manage., 255*(11), 3686–3690.

76. Alcázar, M., Kind, T., Gschaedler, A., Silveria, M., Arrizon, J., Fiehn, O., et al., (2017). Effect of steroidal saponins from *Agave* on the polysaccharide cell wall composition of *Saccharomyces cerevisiae* and *Kluyveromyces marxianus. LWT Food Sci. Technol., 77*, 430–439.

77. Romero-Gonzalez, J., Walton, J. C., Peralta-Videa, J. R., Rodriguez, E., Romero, J., & Gardea-Torresdey, J. L., (2009). Modeling the adsorption of Cr(III) from aqueous solution onto *Agave lechuguilla* biomass: Study of the advective and dispersive transport. *J. Hazard. Mater., 161*(1), 360–365.

78. Hernández-Castillo, F. D., Castillo-Reyes, F., Gallegos-Morales, G., Rodríguez-Herrera, R., & N. Aguilar, C. N., (2011). Plant extracts from Mexican native species: An alternative for Control of plant pathogens. *Research in Organic. Farming. In Tech. 1*, 139–156.

79. Schwarzlin, R., Pu, N., Makuc, D., Kri, M., Vovk, I., Plavec, J., et al., (2016). Synergistic complex from plants *Solanaceae* exhibits cytotoxicity for the human hepatocellular carcinoma cell line HepG2, *BMC Complement Altern. Med.*, *16*(395), 1–12.

80. Lin, C., Lu, W., Wang, C., Chan, Y., & Chen, M., (2013). Capsaicin induces cell cycle arrest and apoptosis in human KB cancer cells. *BMC Complement Altern. Med.*, *13*(1), 13–46

81. Ornelas-Paz, J. D. J., Martínez-Burrola, J. M., Ruiz-Cruz, S., & Santana-Rodríguez, V., (2010). Effect of cooking on the capsaicinoids and phenolics contents of Mexican peppers. *Food Chem.*, *119*(4), 1619–1625.

82. Maksimova, V., Gudeva, L. K., Gulaboski, R., & Nieber, K., (2016). Co-extracted bioactive compounds in *Capsicum* fruit extracts prevent the cytotoxic effects of capsaicin on B104 neuroblastoma cells. *Rev. Bras. Farmacogn.*, *26*(6), 744–750.

83. Sung, J., Bang, M., & Lee, J., (2015). Short communications Bioassay-guided isolation of anti-adipogenic compounds from defatted pepper (*Capsicum annuum* L.) seeds. *J. Funct. Foods.*, *14*, 670–675.

84. Barbieri, R., Coppo, E., Marchese, A., Daglia, M., Sánchez, E., Nabavi, S. F., et al., (2017). Phytochemicals for human disease: An update on plant-derived compounds antibacterial activity. *Microbiol. Res.*, *196*, 44–68.

85. Vieira da Silva, B., Barreira, J. C. M., & Oliveira, M. B. P. P., (2016). Natural phytochemicals and probiotics as bioactive ingredients for functional foods: Extraction, biochemistry and protected-delivery technologies. *Trends Food Sci. Technol.*, *50*, 144–158.

86. Verma, N., & Shukla, S., (2015). Impact of various factors responsible for fluctuation in plant secondary metabolites. *J. Appl. Res. Med. Aromat. Plants.*, *2*, 105–113.

87. Gil, E. S., & Couto, R. O., (2013). Flavonoid electrochemistry: A review on the electroanalytical applications. *Rev. Bras. Farmacogn.*, *23*(3), 542–558.

88. Fu, Z. F., Tu, Z. C., Zhang, L., Wang, H., Wen, Q. H., & Huang, T., (2016). Antioxidant activities and polyphenols of sweet potato (*Ipomoea batatas* L.) leaves extracted with solvents of various polarities. *Food Biosc.*, *15*, 11–18.

89. Martins, N., Barros, L., Ferreira, I. C. F. R., (2016). *In vivo* antioxidant activity of phenolic compounds: Facts and gaps. *Trends Food Sci. Technol.*, *48*, 1–12.

90. De Mello Andrade, J. M., & Fasolo, D., (2013). Chapter 20: Polyphenol antioxidants from natural sources and contribution to health promotion. *Polyphenols in Human Health and Disease Elsevier Inc. Brasil*, (Vol. 1, pp. 253–265).

91. Martins, S., Mussatto, S. I., Martínez-Avila, G., Montañez-Saenz, J., Aguilar, C. N., & Teixeira, J. A., (2011). Bioactive phenolic compounds: Production and extraction by solid-state fermentation. A review. *Biotechnol. Adv.*, *29*(3), 365–373.

92. Qiu, Y., Liu, Q., & Beta, T., (2010). Antioxidant properties of commercial wild rice and analysis of soluble and insoluble phenolic acids. *Food Chem.*, *121*(1), 140–147.

93. Faggio, C., Sureda, A., Morabito, S., Silva, A., Mocan, A., Nabavi, S. F., & Nabavi, S. M., (2017). Flavonoids and platelet aggregation: A brief review. *Eur. J. Pharmacol.*, *807*, 91–101.

94. Sebastian, R. S., Wilkinson, E. C., Goldman, J. D., Steinfeldt, L. C., Martin, C. L., Clemens, J. C., et al., (2017). New, publicly available flavonoid data products: Valuable resources for emerging science. *J. Food Compost. Anal.*, *2017*.

95. Bakhtiari, M., Panahi, Y., Ameli, J., & Darvishi, B., (2017). Protective effects of flavonoids against Alzheimer's disease-related neural dysfunctions. *Biomed. Pharmacother., 93*, 218–229.
96. Raffa, D., Maggio, B., Raimondi, M. V., Plescia, F., & Daidone, G., (2017). Recent discoveries of anticancer flavonoids. *Eur. J. Med. Chem. 142*, 213–228.
97. George, V. C., Dellaire, G., & Rupasinghe, H. P. V., (2017). Plant flavonoids in cancer chemoprevention: Role in genome stability. *J. Nutr. Biochem., 45*, 1–14.
98. Ma, W., Guo, A., Zhang, Y., Wang, H., Liu, Y., & Li, H., (2014). A review on astringency and bitterness perception of tannins in wine. *Trends Food Sci. Technol., 40*(1), 6–19.
99. Barrett, A. H., Farhadi, N. F., & Smith, T. J., (2017). Slowing starch digestion and inhibiting digestive enzyme activity using plant flavanols/ tannins: A review of efficacy and mechanisms. *LWT Food Sci. Technol. 87*, 394–399.
100. Seleem, D., Pardi, V., & Murata, R. M., (2017). Review of flavonoids: A diverse group of natural compounds with anti-*Candida albicans* activity *in vitro. Arch. Oral Biol., 76*, 76–83.
101. Spagnuolo, C., Moccia, S., & Luigi Russo, G., (2017). Anti-inflammatory effects of flavonoids in neurodegenerative disorders. *Eur. J. Med. Chem.*, 1–11.
102. González-Molina, E., Perles, R., Moreno, D. A., & Viguera, C., (2010). Natural bioactive compounds of *Citrus limon* for food and health. *J. Pharm. Biomed. Anal., 51*(2), 327–345.
103. Wang, T., Yang Li, Q., & Bi, K. S., (2017). Bioactive flavonoids in medicinal plants: Structure, activity and biological fate. *Asian J. Pharm. Sci.*, 1–11.
104. Ignant, I., Volf, I., & Popa, V. I., (2011). A critical review of methods for characterization of polyphenolic compounds in fruits and vegetables. *Food Chem., 126*, 1821–1835.
105. Makam, N. S., Chidambara, M. K. N., Sultanpur, C. M., & Rao, R. M., (2014). Natural molecules as tumour inhibitors: Promises and prospects. *J. Herb. Med., 4*(4), 175–187.
106. Paz, J. E. W., Guyot, S., Herrera, R. R., Sánchez, G. G., Juan, C., Esquivel, C., et al., (2013). Alternativas actuales para el manejo sustentable de los residuos de la industria del café en México. Current alternatives for sustainable management of coffee industry by-products in Mexico, *10*, 33–40.
107. Schofield, P., Mbugua, D. M., & Pell, A. N., (2001). Analysis of condensed tannins: A review. *Anim. Feed Sci. Technol., 91*(1–2), 21–40.
108. Arapitsas, P., (2012). Hydrolyzable tannin analysis in food. *Food Chem., 135*(3), 1708–1717.
109. Galea, A. M., & Brown, A. J., (2009). Special relationship between sterols and oxygen: Were sterols an adaptation to aerobic life? *Free Radic. Biol. Med., 47*(6), 880–889.
110. Alvarez, A., Morales, V., Cilla, A., Llatas, G., Siles, L. M., Barberi, R., et al., (2017). Safe intake of a plant sterol-enriched beverage with milk fat globule membrane: Bioaccessibility of sterol oxides during storage. *J. Food Compost. Anal.*, 0–1.
111. Vanmierlo, T., Bogie, J. F. J., Mailleux, J., Vanmol, J., Lütjohann, D., Mulder, M., et al., (2015). Plant sterols: friend or foe in CNS disorders?. *Prog. Lipid Res., 58*, 26–39.

112. Ferrer, A., Altabella, T., Arró, M., & Boronat, A., (2017). Emerging roles for conjugated sterols in plants. *Prog. Lipid Res.*, *67*, 27–37.

113. Gylling, H., Plat, J., Turley, S., Ginsberg, H. N., Ellegard, L., Jessup, W., et al., (2014). Plant sterols and plant stanols in the management of dyslipidaemia and prevention of cardiovascular disease. *Atherosclerosis*, *232*(2), 346–360.

114. Vanmierlo, T., Husche, C., Schött, H. F., Pettersson, H., Lütjohann, D., (2013). Plant sterol oxidation products-Analogs to cholesterol oxidation products from plant origin?. *Biochimie.*, *95*(3), 464–472.

115. Chinsembu, K. C., (2016). Plants and other natural products used in the management of oral infections and improvement of oral health. *Acta. Trop.*, *154*, 6–18.

116. Omonijo, F. A., Ni, L., Gong, J., Wang, Q., Lahaye, L., & Yang, C., (2017). Essential oils as alternatives to antibiotics in swine production. *J. Anim. Physiol. Anim. Nutr. (Berl).*, 1–11.

117. Gadea, R., Glibota, N., Pérez Pulido, R., Gálvez, A., & Ortega, E., (2017). Effects of exposure to biocides on susceptibility to essential oils and chemical preservatives in bacteria from organic foods. *Food Control.*, *80*, 176–182.

118. Raut, J. S., & Karuppayil, S. M., (2014). A status review on the medicinal properties of essential oils. *Ind. Crops Prod.*, *62*, 250–264.

119. Santos, R., Andrade, M., & Sanches-Silva, A., (2017). Application of encapsulated essential oils as antimicrobial agents in food packaging. *Curr. Opin. Food Sci.*, *14*, 78–84.

120. Dar, A. A., & Arumugam, N., (2013). Lignans of sesame: Purification methods, biological activities and biosynthesis – A review. *Bioorg. Chem.*, *50*, 1–10.

121. Tsopmoa, A., Awah, F. M., & Kuetec, V., (2013). Chapter lignans and stilbenes from African medicinal plants. *J. Med. Plants Res.*, Nigeria: Elsevier, 235–277.

122. Gnabre, J., Bates, R., & Huang, R. C., (2015). Creosote bush lignans for human disease treatment and prevention: Perspectives on combination therapy. *J. Tradit. Complement Med.*, *5*(3), 119–126.

123. Kiyama, R., (2016). Biological effects induced by estrogenic activity of lignans. *Trends Food Sci. Technol.*, *54*, 186–196.

124. Slanina, J., & Glatz, Z., (2004). Separation procedures applicable to lignan analysis. *J. Chromatogr. B Analyt. Technol. Biomed. Life Sci.*, *812*(1–2), 215–229.

125. Loypimai, P., Moongngarm, A., & Chottanom, P., (2016). Phytochemicals and antioxidant capacity of natural food colorant prepared from black waxy rice bran. *Food Biosc.*, 15, 34–41.

126. Sparg, S. G., Light, M. E., & Van Staden, J., (2004). Biological activities and distribution of plant saponins. *J. Ethnopharmacol.*, *94*(2–3), 219–243.

127. Vincken, J. P., Heng, L., De Groot, A., & Gruppen, H., (2007). Saponins, classification and occurrence in the plant kingdom. *Phytochemistry*, *68*(3), 275–297.

128. Jacob, J. K., Tiwari, K., Betanzo, J., Misran, J., Chandrasekaran, R., & Paliyath, G., (2012). Biochemical basis for functional ingredient design from fruits. *Annu. Rev. Food Sci. Technol.*, *3*, 79–104.

129. Bacon, K., Boyer, R., Denbow, C., O'Keefe, S., Neilson, A., Williams, R., (2017). Evaluation of different solvents to extract antibacterial compounds from jalapeño peppers. *Food Sci. Nutr.*, *5*(3), 497–503.

130. Guerrouj, K., Sánchez-Rubio, M., Taboada-Rodríguez, A., Roda, R. M., & Iniesta, F., (2016). Sonication at mild temperatures enhances bioactive compounds and microbiological quality of orange juice. *Food Bioprod. Process, 99,* 20–28.

131. Yang, L., Wang, H., Zu, Y., Zhao, Zhang, C., L., Zhang, L., Chen, X., & Zhang, Z., (2011). Ultrasound-assisted extraction of the three terpenoid indole alkaloids vindoline, catharanthine and vinblastine from *Catharanthus roseus* using ionic liquid aqueous solutions. *Chem. Eng. J., 172*(2–3), 705–712.

132. Mohammadhosseini, M., (2017). The ethnobotanical, phytochemical and pharmacological properties and medicinal applications of essential oils and extracts of different Ziziphora species. *Ind. Crops Prod., 105,* 164–192.

133. Girotto, F., Alibardi, L., & Cossu, R., (2015). Food waste generation and industrial uses: A review. *Waste Management, 45,* 32–41.

134. Gonzalez-Sanchez, M. E., Perez-Fabiel, S., Wong-Villarreal, A., Bello-Mendoza, R., & Yanez-Ocampo, G., (2015). Agroindustrial wastes with potential for methane production by anaerobic digestion. *Rev. Argent. Microbiol., 47*(3), 229–235.

135. Cañete A. M., Dueñas, I. M., Hornero, J. E., Ehrenreich, A., Liebl, W., & García, I., (2016). Gluconic acid: properties, production methods and applications: An excellent opportunity for agro-industrial by-products and waste bio-valorization. *Process Biochem., 51*(12), 1891–1903.

136. Barragán, H. B., Díaz, T. A. Y., & Laguna, T. A., (2008). Use of agroindustrial wastes. *Revista Sistemas Ambientales., 2*(1), 44–50.

137. Luchese, C. L., Sperotto, N., Spada, J. C., & Tessaro, I. C., (2017). Effect of blueberry agro-industrial waste addition to corn starch-based films for the production of a pH-indicator film. *Int. J. Biol. Macromol., 104,* 11–18.

138. Basanta, R., Delgado, M. A. G., Martínez, J. E. C., Vázquez, H. M., & Vázquez, G. B., (2007). Sostenibilidad del reciclaje de residuos de la agroindustria azucarera: una revisión. *Cienc. Tecnol. Aliment., 5*(4) 293–305.

139. Pandey, A., Soccol, C. R., Nigam, P., Brand, D., Mohan, R., & Roussos, S., (2000). Biotechnological potential of coee pulp and coee husk for bioprocesses. *Biochem. Eng. J., 6,* 153–162.

140. Corzo, B. D. C., (2012). Evaluación de la actividad antimicrobiana del extracto etanólico. *Rev. Mex. Cienc. Farm., 43*(3), 81–86.

141. Upadhyay, R., Nachiappan, G., & Mishra, H. N., (2015). Ultrasound-assisted extraction of flavonoids and phenolic compounds from *Ocimum tenuiflorum* leaves. *Food Sci. Biotechnol., 24*(6), 1951–1958.

142. Schieber, A., Stintzing, F., & Carle, R., (2001). By-products of plant food processing as a source of functional compounds-recent developments. *Trends Food Sci. Technol., 12,* 401–413.

# CHAPTER 8

# ADVANCES AND OPPORTUNITIES OF ANAEROBIC BIOCONVERSION OF CITRUS WASTE

ALFREDO I. GARCÍA-GALINDO,[1]
MONICA L. CHÁVEZ-GONZÁLEZ,[1]
JANETH MARGARITA VENTURA SOBREVILLA,[1]
RICARDO GÓMEZ-GARCÍA,[1] ROSA SALAS-VALDEZ,[1]
ILEANA MAYELA MORENO DÁVILA,[1]
NAGAMANI BALAGURUSAMY,[2] and CRISTÓBAL N. AGUILAR[1]

[1] Group of Bioprocesses and Bioproducts, Food Research Department, Chemistry School, Autonomous University of Coahuila, Blvd. V. Carranza e Ing. J. Cardenas V., Saltillo, Coahuila, CP 25280, Mexico, Tel.: 52 (844) 416-12-38, Fax: 52 (844) 415-12-38, +52 (844) 415-95-34, E-mail: cristobal.aguilar@uadec.edu.mx

[2] Bioremediation Laboratory, Faculty of Biological Sciences, Autonomous University of Coahuila, Torreón, Coahuila, CP 27000, México

## ABSTRACT

Citrus peel waste is rich in biopolymers and the biomass energy in it can be recovered by anaerobic digestion bioprocess. Nevertheless, the process instability is a common operational issue during anaerobic digestion of citrus wastes. Bioprocess monitoring, control, and microbial management are key points to control instability and increase the energy conversion efficiency of anaerobic digesters. Factors affecting anaerobic digestion process, particularly due to physicochemical properties of the substrate and the formation of undesirable compounds during the process, and their effect on microbial bioconversion efficiency during anaerobic digestion are discussed. Further, advances on anaerobic digestion of citrus wastes are analyzed and discussed in this paper along with perspectives and opportunities for future are presented in this paper.

## 8.1　INTRODUCTION

The agricultural industry generates a great quantity of byproducts with low or no value to the main process, but the biomass energy present in most of them can be harnessed into commercial-valued products [47]. Citric plants belong to the genus *Citrus* [34], and are typically from subtropical and tropical environments with variations in appearance and fruit quality [38].

　　Mexico is 4th bigger producer of citric fruits with the 4.6% of world-wide production, with more than 7.5 millions tons in 2014 [13]. Citrus fruits production in Mexico generated more than 380 millions USD, which supported over than 69 thousand families in rural settlements [45]. Consequently, there is large production of citric wastes after fruit processing and they have a great potential to be used as a substrate for the production of different value-added products, such as dietary fiber, organic acids, essential oils, phenolic compounds, enzymes, biofuels, biopesticides, plastics, fertilizers among others [53].

## 8.2　CLASSIFICATION OF CITRUS FRUITS

Citric fruits belong to *Rutaceae* family and are grouped in *Aurantioideae* subfamily. Most important genera are *Citrus, Poncirus, and Fortuneal*, of which the first mentioned is the most important in agronomic practices representing almost the total production. Classifications proposed by *Swingle* (most important) and Tanaka (complimentary) [38] is followed for present day cultivated *Citrus*, but still the establishment of genetic relationship between *citrus* and allied genera is a riddle [5]. Some of the characteristics of the most cultivated and better-known species of citrus plants are presented in the following subsections.

### 8.2.1   ORANGE (*Citrus sinensis* L. Osb.)

Orange is one of the most popular fruits and Mexico is an important producer worldwide. It is originally from China and obtained from a medium size perennial tree (growth up to 7 m). It can be consumed fresh or processed practically at any time. It can be eaten alone or as part of many dishes [48]. Orange is even used in Mexican traditional medicine against colds, coughs, sore throats, and as antiscorbutic and astringent [3] and have been reported to show antidepressive, anxiolytic [32] and anti-bacterial activity [46].

### 8.2.2   GRAPEFRUIT (*Citrus paradisi* Macf)

Native to the Caribbean, this fruit can be eaten fresh or as part of an elaborated dish. Mexico is the fourth most largest producer of this fruit [13]. The grapefruit is used to produce jellies, juices, and concentrates by the food industry. Grapefruit is a low-calorie food and is a good source of Vitamin C and Inositol. There are two kinds of grapefruit: with white pulp (varieties Dunchan and Marsh) and with pink or red pulp (varieties Star Rubi, Red Blush, etc.) [38]. In recent years, grapefruit extracts are reported to demonstrate regulatory activity over renal failure related-blood pressure [18] and apoptotic activity on leukemia cells [10].

### 8.2.3   MANDARIN (*Citrus reticulata*)

Mandarin shows a wide range of variations in its morphological and molecular characteristics [33]. Different taxonomic systems recognize different species of mandarin: Swingle recognizes 5, Tanaka recognizes 36 and Davies and Albrigo recognize 3 "real" and some others as mandarin hybrids [38]. Mandarin fruits are rich in Vitamin C and essential oils. Its juice is used to add tastes in meat, fish, and shellfish dishes and is also widely used in the preparation of desserts [38]. Chinese traditional medicine is reported to use mandarin for treatment/co-treatment against cancer [28, 42]. Presence of compounds with antifibrotic [63] and bactericidal properties have also been reported [11].

Tangerine is a special case in the family of citrus fruits. Tangerine classification is more difficult due to the heterogeneity of taxonomic systems. In recent years, a karyological separation of mandarin confirmed that tangerine is a subspecies of mandarin, *Citrus tangerina* cv. Dancy [33]. Tangerine well known for the presence of compounds with antioxidant and anti-inflammatory activity, principally in its peel [58, 35].

## 8.2.4   *LEMON (Citrus limon (L.) Burm. F.)*

Lemon is the third most important citrus fruit in the world [13]. It is native to Asia and it grows as a small tree with irregular branches and short thorns. The fruit on reaching maturity stage develops its round shape, rough peel and acid-taste juice. In many earlier cultures, it is used as a remedy for respiratory and nervous illnesses and as antiseptic [3]. Recently, lemon peel extracts have been reported to show anti-inflammatory and anti-oxidant activity [4], to improve the immunological system [19] and to dissolve gallstones [9].

## 8.3   EDAPHIC AND CLIMATOLOGICAL CONDITIONS

### 8.3.1   *WEATHER*

Citrus fruit trees can be raised in a wide range of altitude (from sea level to altitudes higher than 2000 m), and temperature (from 18 to 30°C). The varieties of citrus fruits are chosen depending on their preferred culture conditions and thus can be cultivated under different conditions [38]. In tropical regions, the elapsed time between flower apparition and development of mature orange fruit can be 8–9 months, but in continental weather conditions it can be delayed for 16 months. Climatic conditions influence intensity, duration, and distribution of flowers [38].

Perennial trees, like citrus trees, suffer due to these environmental factors more than other types of cultivars. Principal regions of citrus production have a sub-tropical weather; between 23.5 and 40° latitude in both hemispheres, and this region presents environmental conditions favorable to development, production, and quality of the citrus plants. These factors are also responsible for the great variation in yield of cultivars, which range from 100 ton/ha in sub-tropical locations to 15 ton/ha in tropical regions [38]. Recent changes in precipitation patterns and climate changes affect population and yield of cultivars [54]. Consequently, economic growth of the citrus fruit farmers is greatly affected [45].

## 8.4  CITRUS AGROINDUSTRIAL WASTE

The worldwide production of citrus fruits, including orange, lemons, lime, and mandarins was 139,796,997 tons [14]. Citrus fruit are consumed as fresh fruit or as processed products, like juices, essential oils and marmalades. Apart from the citrus peels, which is the main waste, fruit pulp, fruit kernels, seeds, and membranes are other wastes, and all these wastes constituted about of 50% of the total weight of fruit [50].

Composition of citrus peels waste is characterized by high organic content mainly sugars like as glucose, fructuose, sucrose, pectin, cellulose, and hemicellulose, and low content of lignin and with pH between 3 and 4 [36]. Large quantities of lignocellulosic content of citrus wastes are the most important renewable reservoir of carbon, which can be exploited for variety of vitally important chemical feedstocks and fuels for economic activity and growth of any country.

There are numerous studies on enzymatic hydrolysis of cellulose and hemicellulose present in citrus peel, for the bioconversion of obtained sugars to ethanol by fermentation.

Another alternative use of citrus wastes, to recover the energy in the form of methane by anaerobic digestion, which helps in waste management and as well as energy recovery [43].

## 8.5  FACTORS AFFECTING ANAEROBIC DIGESTION PROCESS

Citrus residues are an excellent option to be employed in anaerobic digestion process, however, several factors can affect the performance of anaerobic digestion process.

## 8.5.1  SUBSTRATE CHARACTERISTICS

### 8.5.1.1  EFFECTS OF PRETREATMENT METHODS OF ORGANIC SOLID WASTE

The important rate-limiting step during anaerobic digestion for complex organic substrates is the hydrolysis step [23], due to physico-chemical factors like particle size, composition, parameters promoting or inhibiting attachment of hydrolytic bacteria on substrate, etc. Pretreatment of substrate is carried out to minimize the limitation due to hydrolysis step. Pretreatment methods can be divided into mechanical, thermal, chemical, biological or a combination of them.

Mechanical pretreatment such as sonication, liquid shear, collision, high-pressure homogenizer, maceration, and liquefaction help to reduce particle size. Thermal pretreatment disintegrated cell membranes, resulting in solubilization of organic compounds [50], solubilization of organic substrates and temperature is reported to have a direct correlation, but it also has been observed that solubilization can be achieved under lower temperatures, but with longer treatment times. In general, chemical pretreatment destroy the organic compounds due to use of strong acids, alkalis or oxidants. However, the degree of effect of chemical pretreatments depend on the method and the properties of substrate. Most of the biological pretreatments include use of specific enzymes (peptidase, carbohydrolase, etc.) and sometimes combination of different enzymes. Many authors have suggested a combination of different methods such as thermo-chemical, thermo-mechanical, etc. for an effective solubilization.

### 8.5.1.2  C/N RATIO

Apart from the organic content of the wastes, the proportion of carbon and nitrogen (C/N) present in the waste is an important parameter that determines the efficiency of the anaerobic digestion process. A substrate with high C/N do not favor an optimal bacterial growth, while substrates with low C/N ratio results in production of ammonia, which is toxic to methanogens and results in the failure of the process. The optimal C/N ratio for anaerobic digestion is to reported to be between 20–35 [59].

## 8.5.2  pH

The operational pH affects the digestion process and the type of products formed during the process. The optimum pH for anaerobic digestion is in the range of 6.8–7.4. The growth rate of microorganisms is affected by pH changing, like methanogenic and acidogenic microorganisms have optimal pH levels. It is reported by some researchers that methanogenesis is favored in the pH range of 6.5–8.2, with optimum at 7.0 and the optimum pH of acidogenesis is between pH 5.5–6.5.

## 8.5.3   ORGANIC LOADING RATE

Organic loading rate represents the amount of volatile solids fed into a digester per day per unit volume under continuous feeding. Volatile solids represent that portion of the organic material that can be digested, while the remainder of the solids is fixed. The fixed solids and a part of the volatile solids are non-biodegradable. The actual loading rates depends on the composition of waste, as the nutrients and their proportion in the wastes determine the level of biochemical activity that occur in the digester.

## 8.5.4   TEMPERATURE

Temperature is one of the most important parameter in anaerobic digestion. It determines the rates of hydrolysis and methanogenesis. Anaerobic digestion is generally applied at two ranges of temperatures: mesophilic (35°C) and thermophilic (55°C). Mesophilic bacteria are more robust and can tolerate greater changes in the environmental conditions. Thermophilic systems offer faster kinetics, higher methane production and pathogen removal; but thermophilic process is more sensitive to toxic substances and is less attractive because it requires more energy consumption.

## 8.5.5   HYDRAULIC RETENTION TIME

Hydraulic retention time is a measure to describe the average time that a substrate resides in a digester. In most cases and if the process is not optimized, shorter HRT results in less efficient degradation of organic matter (as volatile solids or COD), associated with poor process stability.

Hydraulic retention time is defined by the following equation:

$$HRT = \left(\frac{V}{\theta}\right)$$

where V is the biological reactor volume and $\Theta$ the influent flow rate in time.

### 8.5.6 INHIBITORY SUBSTANCES

Inhibition of anaerobic digestion process occurs due to the presence of toxic substances present in the substrate or due to the formation/accumulation of intermediary compounds such as volatile fatty acids, ammonia, hydrogen sulfide, etc. In some cases, continuous accumulation of some of the metabolites mentioned anterior can lead to total failure of the biodigester.

Regarding citrus wastes as a substrate for anaerobic digestion, presence of essential oils in the wastes inhibit the process. It has been reported that essential oils are cytotoxic and is one of the major technical difficulties related to anaerobic digestion of these wastes [44].

#### 8.5.6.1 PRESENCE OF ESSENTIAL OIL AND LIMONENE

In general, essential oils from plants are known for their antimicrobial, antioxidant, antiradical, insecticidal activities [30, 60] and their positive effect in controlling postharvest diseases due to molds have been demonstrated [49]. In addition, essential oils are used extensively as flavoring agents in diverse food production, and in cosmetic industries due its fragrance and flavor [6, 60]. Essential oils of citrus fruits are considered as an important part of their auto-defense mechanism, constituting a barrier against different pathogens [7]. Essential oils are present in the flavedo of peels, is a mixture of terpenes, main monoterpenes, sesquiterpene, hydrocarbons, and oxygenated derivatives such as aldehydes, ketones, esters, alcohols, and acids. About 85–99% of essential oils constitute its volatile fraction and the rest 1–15% are of non-volatile fraction. The concentration of each one of these groups varies according to the type of citrus [1, 15, 52].

The main component of citrus essential oil is limonene (1-methyl-4-(1-methylethenyl)-cyclohexene) represent more than 90% of total content of essential oils) is a cyclic monoterpene with a formula of $C_{10}H_{16}$ [29, 60]. Limonene is a strong antimicrobial effect, and alters the membranes by its incorporation into lipid monolayers, affecting its organization and leading to disruption and leakage of cell components. The effect of this monoterpene is determined by lipid composition of the cellular membrane of the microorganism [20].

Limonene has been shown to inhibit both hydrolytic-acidogenic bacteria and methanogens. Microbial oxidation of carbon double bond (C = C) usually requires the presence of molecular oxygen as a cosubstrate. This is why monoterpenes are considered to be recalcitrant in anoxic media. Branching and terminal ring closures, as in the case of limonene contribute to the molecule stability in anoxic environments. However, the hydrogenation of double bonds in anaerobic media is thermodynamically favorable [44].

Hence, various authors suggest the maximum extraction of limonene from citrus fruits using different kind of methodologies such as; distillation, solid-liquid extraction, solvent extraction and steam explosion [62], so that the inhibition during anaerobic digestion of citrus wastes can be minimized.

## 8.5.6.2   *EXTRACTION METHODS OF LIMONENE*

As mentioned, one of the ways to improve anaerobic digestion of citrus wastes is the prior removal of essential oils from citrus wastes. Two strategies can be applied: (i) recovery of the essential oils to obtain value-added component under biorefinery concept, or (ii) removal of essential oils solely to enhance biogas production [25, 57]. The choice between recovery and removal depends on the economic feasibility of whole process; the former implies a higher cost but yields a valuable product, while the later to involves a cost, but increase in biogas production could compensate the incurred cost. Essential oils from citrus can be extracted by traditional methods such as cold pressing, distillation through the exposure to boiling water or steam and soxhlet system [17]. These conventional methods have some disadvantages related to high-energy costs and long extraction times. Several new methods such as supercritical fluid extraction, ultrasound extraction, controlled pressure drop process, sub-critical water extraction and microwave-assisted extraction have reported to be efficient, time saving and cost-effective extraction process with enhanced yield of essential oils [17, 21, 27]. A spectrum of traditional and non-conventional methods for limonene extraction from citrus wastes is presented in Table 8.1. It can be observed that the extraction strategies for recovery of this value-added product needs to be explored further to improve the yield. Steam explosion has been proposed as pre-treatment to recovery the limonene previous to the anaerobic digestion of citrus wastes and is reported to remove 94.3% of the limonene and increased the stability of thermophilic anaerobic co-digestion of treated citrus wastes with organic fraction [16].

**TABLE 8.1**　Spectrum of Methods Employed for Limonene Extraction From Citrus Wastes

| Citrus waste | Extraction method | Conditions | Yield extraction | References |
|---|---|---|---|---|
| Orange peel | Steam explosion | High-pressure reactor 10 L Temperature 150°C for 20 min Steam pressure 60 bar | 94.3% | [16] |
| Orange peel | Solid-liquid/n-hexane | - | 80% | [55] |
| Lemon peel | Traditional Soxhlet extraction | Matrix/solvent ratio 1:25 Temperature 68°C Time 4 h | 0.95% | [17] |
| Tangerine Peel | Simultaneous steam distillation and solvent extraction | Time 2 h Solvent ethyl alcohol | 7.7% | [39] |
| Lemon | Solvent extraction/ Hexane | Matrix/solvent ratio 1:15 Temperature 50°C Time of exposure 30 min | 2. 97% | [17] |
| Orange Peel | Soxhlet extraction | Temperature 100°C Time 30 min Relation 1:4 w/v | 0.09% | [36] |
| Orange Peel | Supercritical fluid extraction | Time 45 min 318 K 12 MPa | 5.5% | [22] |
| Orange Peel | Simultaneous distillation-extraction | Solvent mixture of n-pentane and diethyl ether (1:1, v/v) | 0.54% | [21] |

## 8.6    APPLICATIONS OF ANAEROBIC DIGESTION

High consumption of fossil fuels for energy and the consequent emission of greenhouse gases, it is necessary to look for alternative clean and sustainable energy production [25]. Recently there are many studies focusing on the use of citrus wastes as biomass energy source employing anaerobic digestion for production of various bio-products and energy recovery (Table 8.2). Anaerobic digestion is well known as cost-effective and technically feasible way to recover the energy present in citrus wastes [61]. Organic matter present in the citrus wastes are degraded under anaerobic conditions by synergistic action of eleven different microbial trophic groups, including bacterial and archaeal domains and the energy is recovered in the form of methane [2,36].

TABLE 8.2    Recent Studies Employing Anaerobic Digestion of Citrus Waste

| Type of reactor | Scale of study | Loading rate | HRT | Product | Yield | References |
|---|---|---|---|---|---|---|
| Semi-continuous | 4 L | 1.91 kg | 8–30 days | Bio-diesel | 0.33 L | [31] |
| Leach bed batch | 5 L | 0.51 kg | 25 days | Methane | 0.39 L | [26] |
| Batch | 0.5 L | - | 7–10 days | Bio-methane | 0.971 g | [8] |
| Batch | 0.120 L | 3 g/day | 30 days | Bio-gas | 0.177 L | [57] |
| Batch | 0.5 L | - | 30 days | Bio-gas | 0.154 L | [12] |
| Batch | 2 L | - | 15 days | Methane | 0.5 L | [37] |
| Semi-continuous | 0.5 L | 1.2 g/L | 70 days | Methane | 0.431 L | [40] |

Martin [31] have shown that the anaerobic digestion of orange peel waste, after d-limonene extraction showed higher methane production, which suggested that this process can be integrated as a biorefinery approach by orange juice processing industries. However, Wikandari [55] observed that methane production was reduced from limonene extracted orange peel due to inhibition of the residual solvents present in the wastes after the solid-liquid extraction with n-hexane.

The use of thermochemical processes, such as incineration, pyrolysis, and gasification, are not recommended as pretreatment methods for use of anaerobic digestion process for energy recovery in the form of methane [24,57] or by fermentation for the production of bio-ethanol [51].

Calabró [8] employed microwave steam diffusion for extraction of essential oils and obtained a yield of 0.971 g/kg TS. They further reported a production of 300 Ln $CH_4$/g VS under thermophilic conditions by co-digestion of extracted orange peel waste with different Italian kitchen wastes.

They further observed that the presence of high concentrations of essential oil in the wastes caused a lag phase in methanogenesis. Previously, Martín [31] extracted 70% of limonene in the wastes by steam distillation method and studied the effect of different metals (Cu, Ni, Zn, Cr, Cd, and PB) on the stability of the anaerobic process in a continuously stirred-tank reactor at 37°C and 67°C. They observed that the rate of biodegradation and methane production was higher in thermophilic reactor than the mesophilic one.

Koppar [26] used a leach bed batch anaerobic digester for biogasification of citrus peel waste under thermophilic (55°C) conditions and reported a cumulative methane yield of 0.644 L $CH_4$ at standard temperature (273 K) and pressure (1 atm) with a HRT of 16 days. They did not observe toxicity due to the presence of essential oils. Later, Wikandari [56] observed that rapid acidification and inhibition by D-limonene are major challenges of biogas production from citrus wastes. They used a membrane bioreactor at 55°C, and with a HRT of 30 days. The hydrolysates formed by free bacteria digesting the citrus wastes passed through the membrane for their conversion into biogas and the methane production was 0.33 $Nm^3$/kg VS, while the conventional anaerobic reactor recorded a methane production of only 0.05 $Nm^3$/kg VS. Approximately 73% of the theoretical methane yield was achieved using the membrane bioreactor.

Pourbafrani [41] developed an integrated process for the production of ethanol, biogas, pectin, and limonene from citrus wastes. Citrus wastes were hydrolyzed using dilute-acid at different temperatures (130, 150 or 170°C) and at different residence times (3, 6 and 9 min) and a maximum yield (0.41 g of sugars/g of the total dry citrus wastes) was obtained at 150°C and at 6 min residence time. The sugars obtained were fermented to ethanol and a yield of 0.43 g/g of the fermentable sugars was reported. Further, they anaerobically digested the stillage and the remaining solid materials of the hydrolyzed citrus wastes for biogas production. They reported that the integrated process resulted in the production of 39.64 L ethanol, 45 $m^3$ methane, 8.9 L limonene, and 38.8 kg pectin from one ton of citrus wastes with 20% dry weight.

## 8.7  CONCLUSIONS

Anaerobic digestion process offers the viable option of recovering energy present in the wastes generated citrus fruit processing industry. But for the process to be efficient and sustainable, it is important to improve the extraction efficiency of essential oils present in these wastes so that their inhibitory effect on microbial groups involved in anaerobic digestion process, in particular, methanogens is minimized to maximize the methane yield. Though diverse strategies exist for extraction, they are not completely efficient and there is a need for developing innovative strategies for higher recovery of limonene, a value-added product from this waste. The bioconversion of citrus wastes for bioenergy production and other different value-added bio-products recovery can aid not only in terms of economic benefits, but also paving pathway in establishment of sustainable citrus processing industries.

## ACKNOWLEDGMENTS

Authors thank CONACYT (National Council of Science and Technology–Mexico) for the financial support.

## KEYWORDS

- anaerobic digestion
- citrus peel waste
- essential oil
- process stability

## REFERENCES

1. Acar, U., Kesbic, O. S., Yilmazm, S., Gültepe, N., & Türker, A., (2015). Evaluation of the effect of essential oil extracted from sweet orange peel (*Citrus sinensis*) on growth rate of tilapia (*Oreochromis mossambicus*) and possible disease resistance against *Streptococcus iniae. Aquaculture*, *437*, 282–286.
2. Alvarado, A., Montañez-Hernández, L. E., Palacio-Molina, S. L., Oropeza-Navarro, R., Luévanos-Escareño, M. P., & Balagurusamy, N., (2014). Microbial trophic inter-

actions and mcrA gene expression in monitoring of anaerobic digesters. *Front Microbiol.*, *5*, 597.

3. Argueta, A., & Vázque, M. C. G., (1994). *Atlas de Las Plantas de La Medicina Tradicional Mexicana.* Ed. Instituto Nacional Indigenista. Mexico.

4. Basli, A.T., Nawel, S., Bachra, K., & Khodira, M., (2016). *In-Vitro* Antioxidant and Anti-Inflammatory Activities of Peel and Peeled Fruits Citrus Limon. *Curr Nutr Food Sci.*, 12(4).

5. Bayer, R, J., Mabberley, D. J., Morton, C., Miller, C. H., Sharma, I. K., Pfeil, B. E., et al., (2009). A molecular phylogeny of the orange subfamily (*Rutaceae: Aurantioideae*) using nine cpDNA sequences. *Am. J. Bot.*, *96*(3), 668–85.

6. Boukroufa, M., Boutekedjiret, C., Petigny, L., Rakotomanomana, N., & Chemat, F., (2015). Bio-refinery of orange peels waste: A new concept based on integrated green and solvent free extraction processes using ultrasound and microwave techniques to obtain essential oil, polyphenols and pectin. *Ultrason Sonochem.*, *24*, 72–79.

7. Caccioni, D. R. L., Guizzardi, M., Biondi, D. M., Renda, A., & Ruberto, G., (1998). Relationship between volatile components of citrus fruit essential oils and antimicrobial action on *Penicillium digitatum* and *Penicillium italicum. Int. J. Food Microbiol.*, *43*, 73–79.

8. Calabró, P. S., Pontoni, L., Porqueddu, I., Greco, R., Pirozzi, F., & Malpei, F., (2016). Effect of the concentration of essential oil on orange peel waste biomethanization: Preliminary batch results. *Waste Mana. G.*, *48*, 440–447.

9. Chekroune, M., & Benamara, S., (2017). Gallstones-dissolving capacity of lemon (*Citrus limon*) juice, *Herniaria hirsuta* L. Extract and lemon juice-based natural vinaigrette *in vitro. Ind. J. Tradit. Know.*, *16*(2).

10. Dai, T., Bo, W., Sheng, L., Jian, J., Li W., & Wen Q., (2016). Pure total flavonoids from *Citrus paradisi* macfad induce leukemia cell apoptosis *in vitro. Chin. J. Integr. Med.*, 310006, 1–6.

11. Damián-Reyna, A. A., González-Hernández, J. C., Maya-Yescas, R., Cortés-Penagos, C. J., & Chávez-Parga, M. C., (2017). Polyphenolic content and bactericidal effect of Mexican *Citrus limetta* and *Citrus reticulata. J. Food Sci. Technol.*, *54*(2), 531–37.

12. Fagbohungbe, M. O., Herbert, B. M. J., Hurst, L., Li, H., Usmani, S. Q., & Semple, K. T., (1016). Impact of biochar on the anaerobic digestion of citrus peel waste. *Biores Technol.*, *216*, 142–149.

13. Food and Agriculture Organization of the United Nations (FAO). Citrus fruits statistics 2015. Rome, Italy.

14. Food and Agriculture Organization of the United Nations (2016), Faostat., http://www.fao.org/faostat/en/#home

15. Fisher, K., & Phillips, C.A., (2006). The effect of lemon, orange and bergamot essential oils and their components on the survival of *Campylobacterjejuni, Escherichia coli* O157, *Listeria monocytogenes, Bacillus cereus* and *Staphylococcus aureusin vitro* and in food systems. *J. Appl. Microbiol.*, *101*(6), 1232–1240.

16. Forgács, G., Pourbafrani, M., Niklasson, C., Taherzadeh, M. J., & Hováth, I. S., (2012). Methane production from citrus wastes: Process development and cost estimation. *J. Chem. Technol. Biotechnol.*, *87*, 250–255.

17. Giovanna, C., Petrillo, F., Alberto, A., Aliakbarian, B., Perego, P., & Calabrò, V. A., (2014). Non-conventional method to extract D-limonene from waste lemon peels and comparison with traditional Soxhlet extraction. *Sep. Purif. Technol.*, *137*, 13–20.

18. Gamboa-Gómez, C., Pérez-Ramírez, I., González-Gallardo, A., Gallegos-Corona, M. A., Ibarra-Alvarado, C., & Reynoso-Camacho, R., (2016). Effect of *Citrus paradisi* and *Ocimum sanctum* infusions on blood pressure regulation and its association with renal alterations in obese rats. *J. Food Biochem.*, *40*(3), 345–57.

19. García-Beltrán, J. M., Espinosa, C., Guardiola, F. A., & Esteban, M. A., (2017). Dietary dehydrated lemon peel improves the immune but not the antioxidant status of gilthead seabream (*Sparus aurata* L.). *Fish Shellfish Immunol.*, *64*, 426–36.

20. Hac-Wydro, K., Flasinski, M., & Romanczuk, K., (2017). Essential oils as food eco-preservatives: Model system studies on the effect of temperature on limonene antibacterial activity. *Food Chem.*, *235*, 127–135.

21. Hong, Y. S., & Kim, K. S., (2016). Determination of the volatile flavor components of orange and grapefruit by simultaneous distillation-extraction. *Korean J. Food Preserv.*, *23*(1), 63–73.

22. Ibrahim, A., & Sarbatly, R., (2012). Effects of modifier polarity on extraction of limonene from *Citrus sinensis* L. osbeck using supercritical carbon dioxide. *Mal. J. Fund. Appl. Sci.*, *8*(2), 115–120.

23. Izumi, K., Okishio, Y. K., Nagao, N., Niwa, C., Yamamoto, S., & Toda, T., (2010). Effects of particle size on anaerobic digestion of food waste. *Int. Biodeterior. Biodegradation.*, *64*(7).

24. Jha, P., & Schmidt, S., (2017). Reappraisal of chemical interference in anaerobic digestion processes. *Renew. Sust. Energ. Rev.*, *75*, 954–971.

25. John, I., Muthukumar, K., Arunagiri, A., & John, I., (2017). A review on the potential of citrus waste for D- Limonene, pectin, and bioethanol production. *Int. J. Green Energy*, 1–14.

26. Koppar, A., & Pullammanappallil, P., (2013). Anaerobic digestion of peel waste and wastewater for on site energy generation in a citrus processing facility. *Energy*, *60*, 62–68.

27. Li, Y., Qi, H., Jin, Y., Tian, X., Sui, L., & Qiu, Y., (2016). Role of ethylene in biosynthetic pathway of related-aroma volatiles derived from amino acids in oriental sweet melons (*Cucumis melo* var. makuwa Makino). *Sci. Hortic.*, *201*, 24–35.

28. Liu, J. Q., & Wu, D.W., (1993). 32 Cases of postoperative osteogenic sarcoma treated by chemotherapy combined with Chinese medicinal herbs. *Chin. J. Integr. Tradit. West Med.*, *13*(3), 132, 150–152.

29. López-Muñoz, G. A., Antonio-Pérez, A., & Díaz-Reyes, J., (2015). Quantification of total pigments in citrus essential oils by thermal wave resonant cavity photopyroelectric spectroscopy. *Food Chem.*, *174*, 104–109.

30. Lou, Z., Chen, J., Yu, F., Wang, H., Kou, X., Ma, C., & Zhu, S., (2017). The antioxidant, antibacterial, antibiofil activity of essential oil from *Citrus medica* L. var. sarcodactylis and its nanoemulsion. *Food Sci. Technol.*, *80*, 371–377.

31. Martín, M. A., Fernández, R., Serrano, A., & Siles, J. A., (2013). Semi-continuous anaerobic co-digestion of orange peel waste and residual glycerol derived from biodiesel manufacturing, *Waste Manag.*, *33*, 1633–1639.

32. Mirghafourvand, M., Sakineh, M., Alizadeh, C., Sevil, H., & Laleh, K., (2017). The effect of orange peel essential oil on postpartum depression and anxiety: A randomized controlled clinical trial. *Iran. Red. Crescent. Med. J.*, *19*(2).

33. Moraes, A. P., Lemos R. R., Brasileiro-Vidal, A. C., Dos Santos-Soares, W., & Guerra, M., (2007). Chromosomal markers distinguish hybrids and non-hybrid accessions of mandarin. *Cytogenet. Genome Res.*, *119*(3–4), 275–81.

34. Morin, C., (1980). *Cultivo de Citricos*. Second Ed. Lima, Perú: Bib. Orton IICA/CATIE.

35. Murakami, A., Nakamura, Y., Ohto, Y., Yano, M., Koshiba, T., Koshimizu, K., et al., (2000). Suppressive effects of citrus fruits on free radical generation and nobiletin, an anti-inflammatory polymethoxy flavonoid. *BioFactors.*, *12*(1–4), 187–92.

36. Negro, V., Mancini, G., Ruggeri, B., & Fino, D., (2016). Citrus waste as feedstock for bio-based products recovery: Review on limonene case study and energy valorization. *Bioresour. Technol.*, *214*, 806–815.

37. Negro, V., Ruggeri, B., & Fino, D., (2017). Recovery of energy from orange peels through anaerobic digestion and pyrolysis processes after d-limonene extraction. *Waste Biomass Valorization, 0*(0), 0.

38. Orduz-Rodríguez, J. O., & Mateus-Cagua, D. M., (2012). Generalities of citrus fruits and agronomic recommendations for their crop in Colombia. In: *Cítrus fruit: Crop, Posharvest and Industrialization*. Ed. Colección Lasallista Investigación y Ciencia, p. 367.

39. Park, S. M., Ko, K. Y., & Kim, I. H., (2015). Optimization of d-limonene Extraction from tangerine peel in various solvents by using soxhlet extractor. *Korean Chem. Eng. Res.*, 717–722.

40. Pellera, F., & Gidarakos, E., (2017). Anaerobic digestion of solid agroindustrial waste in semi-continuous mode: Evaluation of mono-digestion and co-digestion systems. *Waste Manag.*, *68*, 103–119.

41. Pourbafrani, M., Forgács, G., Horváth, I. S., Niklasson, C., & Taherzadeh, M. J., (2010). Production of biofuels, limonene and pectin from citrus wastes. *Bioresour. Technol.*, *101*(11), 4246–4250.

42. Qian, S. H., Wang, Y. X., Yang, N. Y., & Yuan, L. H., (1003). Study on the anticancer activities (*in vivo*) of the extract from citrus reticulata blanco and its influence on sarcoma-180 cells cycle. *Zhongguo Zhong Yao Za Zhi.*, *28*(12), 1167–70.

43. Rezzadori, K., Benedetti, S., & Amante, E. R., (2012). Proposals for the residues recovery: Orange waste as raw material for new products. *Food Bioprod. Process, 90*, 606–614.

44. Ruiz, B., & Flotats, X., (2014). Citrus essential oils and their influence on the anaerobic digestion process: An overview. *Waste Manag.*, *34*(11), 2063–2079.

45. Sagarpa (2012). México, Entre Los Líderes En Producción de Cítricos a Nivel Mundial. Mexico.

46. Sánchez-Recillas, A., Arroyo-Herrera, A. L., Araujo-León, J. A., Hernández-Núñez, E., & Ortiz-Andrade, R., (2017). Spasmolytic and antibacterial activity of two *citrus Sinensis* osbeck varieties cultivated in Mexico. *Evid Based Complement Alternat. Med.*, 1–7.

47. Saval, S., (2012). Aprovechamiento de residuos agroindustriales: Pasado, presente y futuro. *BioTecnologia., 16*(2), 14–46.

48. SIAP (2010–2012). Padrón de productores de cítricos con predios georeferenciados tamaulipas. Ciudad Victoria, Tamaulipas.

49. Simas, D. L. R., De Amorim, S. H. B. M., Goulart, F. R. V., Alviano, C. S., Alviano, D. S., & DaSilva, A. J. R., (2017). Citrus species essential oils and their components can inhibit or stimulate fungal growth in fruit. *Ind. Crops Prod., 98*, 108–115.

50. Su, H., Tan, F., & Xu, Y., (2016). Enhancement of biogás and methanization of citrus waste via biodegradation pretreatment and subsequent optimized fermentation. *Fuel., 181*, 843–851.

51. Taghizadeh-alisaraei, A., Hosseini, S. H., Ghobadian, B., & Motevali, A., (2017). Biofuel production from citrus wastes: A feasibility study in Iran. *Renew. Sustainable Energy Rev., 69*, 1100–1112.

52. Tao, N., Jia, L., & Zhou, H., (2014). Anti-fungal activity of citrus reticulate blanco essential oil against *Penicillium italicum* and *Penicillium digitatum. Food Chem., 153*, 265–271.

53. Wadhwa, M., Bashki, M. P. S., & Makkar, H. P. S., (2015). Wastes to worth: Value-added products from the fruit and Vegetable wastes. *CAB Reviews, 10*(43), 1–25.

54. Walter, J., Harter, D. E. V., Beierkuhnlein, C., Jentsch, A., & De Kroon, H., (2016). Transgenerational effects of extreme weather: Perennial plant offspring show modified germination, growth and stoichiometry. *J. Ecol., 104*(4), 1032–1040.

55. Wikandari, R., Gudipudi, S., Pandiyan, I., Millati, R., & Taherzadeh, M. J., (2013). Inhibitory effects of fruit flavors on methane production during anaerobic digestion. *Bioresour. Technol., 145*, 188–192.

56. Wikandari, R., Millati, R., Cahyanto, M. N., & Taherzadeh, M. J., (2014). Biogas production from citrus waste by membrane bioreactor. *Membranes, 4*(3), 596–607.

57. Wikandari, R., Nguyen, H., Millati, R., Niklasson, C., & Taherzadeh, M. J., (2015). Improvement of biogas production from orange peel waste by leaching of limonene, *Biomed. Res. Int.*, p. 6.

58. Wang, W., & Chen, W. W., (1991). Antioxidative activity studies on the meaning of same original of herbal drug and food. *Zhong Xi Yi Jie He Za Zhi., 11*(3), *134,* 159–161.

59. Yang, X., Choi, H. S., Park, C., & Kim, S. W., (2015). Current states and prospects of organic waste utilization for biorefineries. *Renew. Sustainable Energy Rev., 49*, 335–349.

60. Zarrad, K., Hamouda, A. B., Chaiel, I., Laarif, A., & Jemaa, J. M. B., (2015). Chemical composition, fumigant and anti-acetylcholinsterase activity of the Tunisian *Citrus aurantium* L. essential oils. *Ind. Crops Prod., 76*, 121–127.

61. Zhang, H., Xu, J., Su, X., Bao, J., Wang, K., & Mao, Z., (2017). Citric acid production by recycling its wastewater treated with anaerobic digestion and nano filtration. *Process Biochem., 58*, 245–251.

62. Zhao, C., Zou, Z., Li, J., Jia, H., Liesche, J., Fang, H., & Chen, S., (2017). A novel and efficient bioprocess from steam exploded corn stover to ethanol in the context of onsite cellulase production. *Energy, 123*, 499–510.

63. Zhou Xian-Mei, Zhen-Dong, C., Na, X., Qi, S., & Jian-Xin L., (2016). Inhibitory effects of amines from citrus reticulata on bleomycin-induced pulmonary fibrosis in rats. *Int. J. Mol. Med.*, *37*(2), 339–346.

# CHAPTER 9

# A COMPREHENSIVE REVIEW ON ESSENTIAL OIL NANOEMULSIONS AS AN ALTERNATIVE TO CONTROL MICROBIAL PATHOGENICITY

J. S. SWATHY, AMITAVA MUKHERJEE, and NATARAJAN CHANDRASEKARAN

*Centre for Nanobiotechnology, VIT University, Vellore–632014, Tamil Nadu, India, Tel.: +91-416-220262, E-mail: nchandrasekaran@vit.ac.in*

## ABSTRACT

Aquaculture is the important sector, which provides nutritional security and employment. Disease outbreak is one of the major constraints to the socio-economic development in many countries. The management of healthy fishes and controlling infection leads to the development of drugs. Wide usage of antimicrobial drugs leads to the accumulation of resistant strain in the fish bodies and negative impact on the environment. Several alternative methods have been developed in order to overcome the negative impact of drug usage. Biological control strategies have been developed with less impact on humans and environment, which helps in the reduction of disease. Recently, increasing attention is made on the use of essential oil nanoemulsion as an alternative to existing antimicrobial agents for controlling infections. The oil phase of nanoemulsion consists of essential oil with active ingredients, which is responsible for all biological activities. This article mainly aims to give an overview about the development of alternative method to control the pathogenic infection and their application in aquaculture.

## 9.1   INTRODUCTION

Globally, India is the second important country in the case of annual fisheries and aquaculture production. Aquaculture is one of important sector, which supplies good nutritional security, development in exporting field and also giving new sources of employment to people in rural areas. Disease becomes a one of the main reason for the loss in total aquaculture production and it has an indirect influence in the economic and social development of our country. Climate change favored the outbreak of various pathogenic diseases and its growth. Alternative therapeutic strategies were developed in order to handle the pathogenic disease caused by viruses, bacteria, fungi, and other emerging pathogens. Non-infectious disease also occurred in aquaculture due to the stress, poor management, water quality, lack of nutrition, degradation in aquatic environment, and exposure to contamination, which increases the mortality rate compared to others [1].

Main causative organism for the fish diseases includes parasites, fungi [2, 3], bacteria [4–6] and viruses [7, 8]. Symptoms mainly include tail-fin rots, Ulcers, Cloudy eye, Dropsy, White spot, Finrot, Swimbladder disorder and Lymphocystis. Among all the causative agents for infection, Bacteria play an important role in mortality increase. Bacterial infection mainly consists of (i) gram-negative bacteria that includes *Aeromonas, Pseudomonas, Vibrio, Edwardsville, Flavobacterium, Francisella, Photobacterium, Piscirickettsia, Tenacibaculum,* and *Yersini* and (ii) Gram-positive bacteria includes *Lactococcus, Renibacterium,* and *Streptococcus* [4].

Traditional method to control infections, mainly includes usage of antibiotic, which made a great reduction in mortality [9, 10]. Main aim of antibiotic usage is to improve the growth performances of cultured animal, which was limited due to the occurrences of resistant strains and bioaccumulation in environment and non-targeted species [11, 12]. Antibiotic resistance becomes a major issue which leads to the development of resistance strains that remains in fish tissue which finally transfers to humans. Another side effects include the reduction of larval growth, and inhibit the defense mechanism. As the immune systems becomes weaken it is difficult to respond and this leads to the increase in mortality rate. Moreover, the farmers are not aware about the side effects of antibiotic usage. They used to handle antibiotics without proper safety measure, which leads to skin allergy, redness (dermatitis) [13], aplastic anemia and other severe health issues. Bioaccumulation, is considered as another major problem faced due to the over usage of antibiotics. About 80% of antibiotics are released into the environment, in which most of the antibiotics are found to be persistent and others are easily degradable, exposure to aquatic environment [14–17]. Nowadays, most of the countries banned the use of certain antibiotics due to the health concern and biosafety of non-targeted species [18]. Farmers and hatchery operators also trained in order to handle the antibiotics properly due to its negative impacts in human health and environment.

Vaccination also considered as an alternative prophylactic method for prevention of disease. From past few decades, certain vaccines were developed for controlling infectious disease, but this not at all suggested due to time-consuming, expensive, and stressful to the fishes [19]. Probiotics was considered as another attractive approach, which can be used as a dietary supplement that helps to increase the health status of aquatic animals [20]. Most probiotic belongs to the families such as vibrionaceae, lactic acid bacteria, bacillus species, yeasts, nitrosomonas, nitrobacteria, and sulphide oxidizers. This situation made a way for the development of an effective antibacterial agent using biological strategies leads to reduction in disease outbreak and increases the biosafety. Biological strategies mean the control of disease using a natural biological process or product of the natural process. [21]. Reasons behind the usage of biocontrol therapy in order to control infection includes: safety to aquatic environment and non-targeted species, low cost and easy production, long shell life and stability and easy to handle. Several studies showed the bioactivities of natural products from plants, fungus, and algae, which was found to be of great interest in the prevention or treatment of pathogens [22–24].

Essential oils were produced by aromatic and medicinal plants as secondary metabolites, which plays an important role in the protection of plant. Generally essential oil constitutes of two major groups, one group comprises of terpene and terpenoids, whereas other constitutes of aromatic and aliphatic components. So these oils have used for the treatment of infectious disease, which can be administered orally, topically or aromatherapy. Several studies [25–28] reported about the antimicrobial properties of essential oils which leads to discovery of novel antimicrobial agent to treat pathogenic infections. Even though, essential oils have many therapeutic properties, but there is some restriction due to their insolubility nature. To overcome this problem, nanoemulsion prepared using essential oils can act as an alternative strategy as well as eco-friendly method with less toxicity for controlling the disease outbreak. This review mainly focuses to give an overview about the usage of Essential oil nanoemulsion as an antimicrobial agent.

## 9.2    BIOACTIVITY OF ESSENTIAL OILS AND THEIR CONSTITUENTS

Since ancient times, essential oil has been used for medical treatments, as preservatives and flavoring agent in food. Later on, usage of essential oil such as jasmine, lavender in daily life become common, which gives mental relaxation and improve the life qualities. Generally essential oils found in secretary cavities, glandular trichomes and epidermal cells of aromatic plants [29]. Essential oils are obtained from the different parts of aromatic plants, which are low molecular weight and lipid soluble in nature. Essential oils are considered as secondary metabolites which doesn't involves directly in growth and development. It only involves in defense mechanism which results in the antibacterial activity. Different methods are used to obtain essential oil from aromatic plants, which mainly includes hydrodistillation, solvent extraction, supercritical fluid extraction and cold pressing [30–36] Compared to all the techniques, steam distillation method is widely used method used for the commercial purpose and the extracted oils shows higher antibacterial activity and have natural organoleptic characteristics [34]. Depending on the climate soil composition and vegetative cycle, physicochemical properties of essential oil varies [33]. Samples collected after flowering time tends to give more antibacterial activity compared to other times.

Chemical composition of essential oil is more complex, and it exhibit higher bioactivity when it is in oxygenated or active state. Mainly composed of terpenes, terpenoids, and other aromatic and aliphatic constituent with low molecular weight. Generally, Terpenes/Terpenoids synthesized by mevalonic pathway inside the cytoplasm of the cell and it is composed of isoprene units $(C_5H_8)n$ [37]. Terpenes are mainly classified into monoterpenes $(C_{10}H_{16})$, sesquiterpenes $(C_{15}H_{24})$, diterpenes $(C_{20}H_{32})$, and triterpenes $(C_{30}H_{40})$. Bioactive compounds mainly constitute of monoterpenes such as $p$-cymene, limonene, $\alpha$-pinene, $\alpha$-terpinene, camphor, carvacrol, eugenol, and thymol, diterpenes such as cembrene C, kaurene, and camphorene and sesquiterpene such as caryophyllene, germacrene D, spathulenol, caryophyllene oxide and humulene [35, 38– 41] Terpenoids are group of terpenes which add oxygen molecules or removing their methyl group. Antibacterial activity of these terpenoids depends upon their functional group and hydroxyl group of terpenoids [42–44]. Phenylpropenes, small part of essential oils which was produced during the first step of phenylpropanoid biosynthesis. Most of the molecule such as Eugenol, isoeugenol, vanillin, safrole, and cinnamaldehyde exhibis high antibacterial activity due to the presences of free radical [45–49].

Compared to Gram-positive bacteria's, Gram-negative bacteria's are more resistant to essential oils due to the complexity in cell structure. Depending on the chemical structure of active components and type of bacteria, mode of action and antibacterial activity varies. The hydrophobic nature of Eos, which helps to penetrate inside microbial cells and cause variation in its structure and functionality. Cell membrane is an essential for bacterial survival, in which all the biological activities taking place. Membrane act as an effective barrier between cytoplasm and environment, helps in the exchange of essential ions and metabolites [50]. As the essential oil enters the cell membrane, proteins present in membrane got degraded this leads to the increase in leakage of cell contents [51], reduce the ATP synthesis and finally result in cell lysis and death figure 9.1 [52, 53]. In some other cases, interaction with essential oil leads to the destabilization of phospholipid bilayer, plasma membrane and composition, loss of intracellular components, alteration in electron transport system and inactivation of enzymatic mechanisms [54–57].

Several work has been reported on the antimicrobial activity of plant extract against Gram-negative and Gram-positive bacteria's [58]. Aqueous extract of F. vesca (leaves and fruits) exhibits antibacterial activity against A. hydrophila, Y. ruckeri, and V. anguillarum. Essential oil and methanolic extract of M. longifolia and P. russeliana have been reported as some antimicrobial agents against S. aureus, E. coli, Listeria monocytogenes, K. pneumoniae, A. hydrophila, B. cerus, S. typhimurium, P. aeruginosa and also inhibited the growth of V. anguillarum and S. agalactiate [58–60]. The extract from Piper betle, Syzygium aromaticum, Phyllanthus niruri exhibited a strong antimicrobial activity due to the presences of the active components such as sterol, hydroxychavicol, eugenol, and phenolic compounds [61, 62]. Similarly, several reports have been published using nanoemulsion as an alternative measure for controlling antimicrobial activities, which was prepared using various essential oils. Noguchi et al. [63], Thomas et al., [64], Mishra et al., [65], and Thomas et al., [66] described the usage of nanoemulsion in aquaculture industries such as neem oil nanoemulsion exhibit antibacterial efficacy against pathogenic bacterial species like *Aeromomas salmonicida* and *Pseudomonas aeruginosa*, Lime oil nanoemulsion also exhibits antibacterial efficiency against prominent pathogenic bacteria such as *Pseudomonas aeruginosa*. Another nanoemulsions prepared using essential oils such as cinnamon oil [67–70], Basil oil [67], eucalyptus oil [71, 72], carvacrol and peanut oil [73], d-limonene and sunflower oil [74–76], cinnamon bark [77], clove [78–82], citronella [83, 84], citral essential oil [85], lemograss oil [86, 87], cymbopogon citratus [88], patchouli oil [89], black cumin oil [90–92], tea tree oil [74, 93], and thyme oil [94–95] showed good antibacterial activity against various pathogens [96].

## 9.3 NANOEMULSION AS AN EFFECTIVE MEASURE FOR CONTROLLING MICROBIAL INFECTION

### 9.3.1 NANOEMULSION: ITS PREPARATION AND CHARACTERIZATION

Nanoemulsions are considered as the class of emulsions with a droplet diameter of less than one micrometer (20–200 nm) which has low viscosity

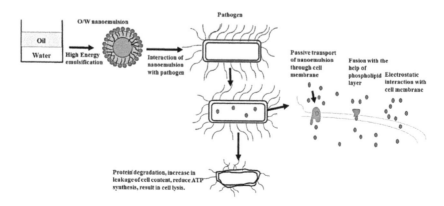

**FIGURE 9.1** Schematic representation of antibacterial mechanism exhibited by nanoemulsion system.

and transparent or translucent in nature [97–101]. Normally, nanoemulsions were considered as kinetically stable, which consist of two immiscible phases such as oil and aqueous phase [100, 102]. The immiscible nature is due to the surface tension present between both phase. Surfactant helps to reduce the interfacial surface tension and result in the formation of a homogenous colloidal system.

Most of the surfactant used is amphiphilic in nature, which contains water-soluble and water-insoluble part. Surfactants mainly classified into four categories such as anionic, cationic, non-ionic, and zwitterionic. Surfactant or an emulsifier was choosing on the basis of following criteria's such as (i) Easy adsorption to two different interfacial surfaces (oil/water) during homogenization, (ii) Reduction in the interfacial surface tension, (iii) stabilization between two phases by forming an interfacial membrane with the help of steric or electrostatic interactions between droplets. Hydrophile- lipophile balance (HLB) is considered as one of the important system that has the tendency to disperse it in polar or non-polar liquid [103, 104]. Co-surfactants also help to penetrate into surfactant molecule and further leads to the reduction of surface tension [105]. Nanoemulsion can be classified on the basis of composition and morphology (i) Oil/Water nanoemulsion: Emulsion which have oil as dispersed phase and water as the continuous phase, (ii) Water/Oil nanoemulsion: Emulsion which have water as dispersed phase and oil as the continuous phase, (iii) Bi-continuous nanoemulsion: This type of nanoemulsion system comprises of micro-domains of oil and water are inter-dispersed within the system like w/o/w/o or o/w/o/w.

When an applicative part is taken into account, stability is considered as one of the important parameter. Main instability phenomena's observed in a nanoemulsion system includes coalescence, Ostwald ripening, and flocculation [106]. Generally, nanoemulsion are resistant to the instability phenomena's such as creaming and sedimentation because their Brownian motion that is sufficient to overcome the force of gravity separation and flocculation [107–111]. Coalescence happens due to the rupturing of films in continuous phase, which result in the fusion of two droplets together to form a single larger droplet. Ostwald ripening phenomena consist of a diffusive transfer of the dispersed phase from the smaller to larger droplets, which leads to the condensation of all droplets into a single one i.e. as the rate of growth decreases with increase of droplet size [112–114]. The

Lifshitz- Slezov and wagner (LSW) theory says that the droplets of the dispersed phase are spherical, and the distances between them are higher than the droplet diameter [115].

Nanoemulsion cannot be formed spontaneously, some amount of energy is needed. Most commonly used strategies for preparation includes low and high energy for emulsification [116–118]. The high-energy methods such as high-pressure homogenizers or microfluidizers and Ultrasonicator, which uses mechanical shear stress to break up the oil and water phases to form nanosized droplets [83, 84, 119, 120]. In this emulsification method droplet size of nanoemulsion can be controlled and varied by depending upon the sample composition, type of instrument and its operating time, pressure, temperature. Ultrasonication is one of the high-energy method, which use ultrasonic waves for breaking the droplets. Parameters which influence the formulation of the nanoemulsion includes surfactant concentration and the sonication time. When the tip of the sonotrode comes into contact the sample generates mechanical vibration and leads to the formation of vapor cavities [121]. Cavity formation and collapse of microbubbles by the pressure fluctuations of a simple sound wave, which forms the extreme levels of highly localized turbulence. Therefore, the turbulent micro-implosions help in breaking the primary droplets into submicron size [122–124]. High-pressure homogenization is considered as a high-energy method in which high-pressure is applied to the system which results in the formation of nanoemulsion with low droplet size. The prepared coarse emulsion was pumped through the inlet valve and it undergo a combination of intense disruptive forces such as turbulence, hydraulic shear, and cavitation, act together to break down larger droplets into smaller ones [125]. Droplet size of the nanoemulsion prepared using this method mainly depends on the homogenizing pressure, cycle, and time affect [108]. Microfluidizer is the one of the high-energy approach for nanoemulsion preparation, which use micro-fluidization technology and high-pressure positive displacement pump (5000–30,000 PSI). The coarse emulsion is passed through the interaction chamber that consists of a microchannel, which results in the formation of fine particles of submicron range [122, 126, 127]. Different type of interaction chamber is used for different application. The Y-type chamber is used in nanoemulsion application purpose. The high-energy method is widely used in cosmetic and pharmaceutical industry [128].

The low energy methods consist of PIT (Phase inversion temperature) method, emulsion inversion point (EIP) and the spontaneous emulsification [129]. In low-energy emulsification method, the stored energy present in the system helps to form small droplets which can be achieved by changing the factors such as temperature, composition that affects the HLB of the systems. [117, 130] Low-energy method is considered as simple and cheaper method for the formulation of nanoemulsion. Spontaneous emulsification also called solvent displacement or self-emulsification, in which spontaneously formulated through mixing the two different phases (oil and aqueous phase) by varying composition or environment [111, 131, 132]. This method is widely used in pharmaceutical industry for the preparation of drug delivery systems such as self-emulsifying drug delivery systems (SEDDS) or Self-Nano Emulsifying Drug Delivery Systems (SNEDDS). The main limitation of this method in the food industry was the usage of high surfactant concentration. [133]. Other low-energy methods occur due to temperature variations is known as PIT – Phase Inversion Temperature. Nanoemulsion can be formed spontaneously with the help of change in temperature-time profile of the certain mixture of oil, water, and the non-ionic surfactant [134–136]. Inversion can be induced either by altering the physicochemical factors such as temperature and electrolyte concentration that indeed affect the HLB of the surfactant in the nanoemulsion system. Phase inversion can also occur at the constant temperature by changing the HLB value of surfactant using a different mixture of surfactants [102, 137]. Phase Inversion Composition (EPI/PIC) is similar to the phase inversion temperature method. In this method instead of temperature, change in the composition of the surfactant occurs which alter the curvature of the surfactant [125, 137].

Characterization of formulated nanoemulsions includes both the physical and chemical assessments which mainly consists of the nanoemulsion components compatibility, isotropicity, and uniformity of the formed nanoemulsion, droplet size, pH, conductivity, viscosity, surface tension, and zeta potential of the dispersed phase [138–141]. Surface morphology was carried out using both transmission electron microscopy (TEM) and scanning electron microscopy (SEM). SEM helps in obtaining the 3-dimensional image of the globules [142]. A proper exploration of surface morphology of disperse phase in the nanoemulsion formulation can be obtained and the image analysis software helps to obtain result of the shape and surface morphology [143]. Qualitative measurements of size and its distribution of TEM micrographs can be accomplished using a digital image-processing programme [144]. To explore the structure and behavior of formation nanoemulsion, sophisticated techniques such as x-ray or neutron scattering, cryo-electron microscopy, atomic force microscopy can be used. The droplet size measurement of the emulsion system is an essential factor in self-nano-emulsification performance as it helps in determinating the release kinetics as well as its absorption. Photon correlation spectroscopy (PCS) and light scattering techniques like static light scattering (SLS), dynamic light scattering (DLS) are a useful method for determination of nanoemulsion droplet size [145]. Conductivity, viscosity measurements helps in providing the useful information at the macroscopic level [146–148]. Turbidity measurements of the nanoemulsion are carried out to estimate the rapid equilibrium that is achieved by the dispersion and reproducibility of this process [142].

## 9.4   CONCLUSION

Recent studies show that nanoemulsion prepared using essential oil have begun to receive more attention due to its safe nature and it became one of the major growth areas in this scientific circle. Traditional methods result in so many toxic behaviors, which threatens both human as well as the environment. Essential oil nanoemulsion helps to replace the existing antimicrobial agent and act as a natural alternative medicine due to the presence of bioactive compounds.

## KEYWORDS

- antibiotics
- antimicrobial activity
- aquaculture
- bacterial resistant
- essential oils
- nanoemulsion

# REFERENCES

1. Subasinghe, R., (2009). Disease control in aquaculture and the responsible use of veterinary drugs and vaccines: The issues, prospects and challenges. *Options Méditerranéennes, 86*, 5–11.

2. Ramaiah, N., (2006). A review on fungal diseases of algae, marine fishes, shrimps and corals. *Indian J. Mar. Sci., 35*, 380–7.

3. Khoo, L., (2000). Fungal diseases in fish. *Seminars in Avian and Exotic Pet Medicine, 9*, 102

4. Frans, I., Michiels, C. W., Bossier, P., Willems, K. A., Lievens, B., Rediers, H., et al., (2011). Vibrio anguillarum as a fish pathogen: Virulence factors, diagnosis and prevention, *J Fish Dis., 34*, 643–61.

5. Wang, W., (2011). Bacterial diseases of crabs: A review. *J. Invertebr. Pathol., 106*, 18–26.

6. Birkbeck, T. H., Feist, S. W, Verner-Jeffreys, D. W., et al., (2011). Francisella infection in fish and infections in fish and shellfish. *J. Fish Dis., 34*, 173–87

7. Guo, F. C., Woo, P. T., et al., (2009) Selected parasitosis in cultured and wild fish. *Vet Parasitol., 163*, 207–216.

8. Brooker, A, J., Shinn, A. P., Bron, J. E., et al., (2007). A review of the biology of the parasitic copepod *Lernaeocera branchialis* (L., 1767) (Copepoda: Pennellidae).

9. Caballero, B., Trugo, L. C., Finglas, P. M., et al., (2003). *Encyclopaedia of Food Sciences and Nutrition.* Amsterdam: Academic Press.

10. Serrano, P. H., (2005). *Responsible Use of Antibiotics in Aquaculture* (No. 469). Food & Agriculture Org.

11. Smith, P., Hiney, M. P., & Samuelsen, O. B., (1994). Bacterial resistance to antimicrobial agents used in fish farming: a critical evaluation of method and meaning. *Annual Review of Fish Diseases, 4*, 273–313.

12. Smith, V. J., Desbois, A. P., & Dyrynda, E. A., (2010). Conventional and unconventional antimicrobials from fish, marine invertebrates and micro-algae. *Marine Drugs, 8*(4), 1213–1262.

13. Rice, R. H., Cohen, D. E., et al., (1996). Toxic response of the skin. In: *Casarett and Doull's Toxicology, The Basic Sciences of Poisons* (5th edn., pp. 529–546). New York: McGraw Hill.

14. Lim, S. J., Jang, E., Lee, S. H., Yoo, B. H., Kim, S. K., & Kim, T. H., (2013). Antibiotic resistance in bacteria isolated from freshwater aquacultures and prediction of the persistence and toxicity of antimicrobials in the aquatic environment. *Journal of Environmental Science and Health, Part B., 48*(6), 495–504.

15. Samuelsen, O. B., (1989). Degradation of oxytetracycline in seawater at two different temperatures and light intensities, and the persistence of oxytetracycline in the sediment from a fish farm. *Aquaculture, 83*, 7–16

16. Weston, D. P., (1995). Environmental considerations in the use of antibacterial drugs in aquaculture. In: Baird, D. J., Beveridge, M. C. M., Kelly, L. A., & Muir, J. F., (eds.), *Aquaculture and Water Resource Management* (pp. 140–165). Fishing News Books. Blackwell, Oxford.

17. Holmström, K., Gräslund, S., Wahlström, A., Poungshompoo, S., Bengtsson, B. E., & Kautsky, N., (2003). Antibiotic use in shrimp farming and implications for environmental impacts and human health. *International Journal of Food Science & Technology, 38*(3), 255–266.

18. Lee, S., Najiah, M., Wendy, W., & Nadirah, M., (2009). Chemical composition and antimicrobial activity of the essential oil of *Syzygium aromaticum* flower bud (Clove) against fish systemic bacteria isolated from aquaculture sites. *Frontiers of Agriculture in China, 3*(3), 332–336.

19. Midtlyng, P. J., Reitan, L. J., Lillehaug, A., & Ramstad, A., (1996). Protection, immune responses and side effects in Atlantic salmon (*Salmo salar* L.) vaccinated against furunculosis by different procedures. *Fish Shellfish Immunol., 6*, 599–613.

20. Verschuere, L., Rombaut, G., Sorgeloos, P., & Verstraete, W., (2000). Probiotic bacteria as biological control agents in aquaculture. *Microbiology and Molecular Biology Reviews, 64*(4), 655–671.

21. Wilson, C. L., (1997). Biological control and plant diseases-a new paradigm. *Journal of Industrial Microbiology & Biotechnology, 19*(3), 158–159.

22. Tagboto, S., & Townson, S., (2001). Antiparasitic properties of medicinal plants and other naturally occurring products. *Advances in Parasitology, 50*, 199–295.

23. Palavesam, A., Sheeja, L, & Immanuel, G., (2006). Antimicrobial properties of medicinal herbal extracts against pathogenic bacteria isolated from the infected grouper *Epinephelus tauvina*. *Journal of Biological Research, 6*, 167–176.

24. Perumal Samy, R., & Gopalakrishnakone, P. (2010). Therapeutic potential of plants as anti-microbials for drug discovery. *Evidence-Based Complementary and Alternative Medicine, 7*(3), 283–294.

25. Akthar, M. S., Degaga, B., & Azam, T., (2014). Antimicrobial activity of essential oils extracted from medicinal plants against the pathogenic microorganisms: A review. *Issues in Biological Sciences and Pharmaceutical Research, 2*(1), 1–7.

26. Duschatzky, C. B., Possetto, M. L., Talarico, L. B., García, C. C., Michis, F., Almeida, N. V., et al., (2005). Evaluation of chemical and antiviral properties of essential oils from South American plants. *Antiviral Chemistry and Chemotherapy, 16*(4), 247–251.

27. Hammer, K. A., Carson, C. F., & Riley, T. V., (1999). Antimicrobial activity of essential oils and other plant extracts. *Journal of Applied Microbiology, 86*(6), 985–990.

28. Al-Mariri, A., & Safi, M., (2014). *In vitro* antibacterial activity of several plant extracts and oils against some gram-negative bacteria. *Iranian Journal of Medical Sciences, 39*(1), 36.

29. Arumugam, G., Swamy, M. K., & Sinniah, U. R., (2016). *Plectranthus amboinicus* (Lour.) Spreng: botanical, phytochemical, pharmacological and nutritional significance. *Molecules, 21*(4), 369.

30. Billot, M., & Wells, F., (1975). *Perfumery Technology: Art.* Science, Industry. John Wiley, New York 1. 462 p.

31. Handa, S. S., Khanuja, S. P., & Longo, G., (2008). Extraction technologies for medicinal and aromatic plants. *International Centre for Science and High Technology, 3*(11), 21–25. 266 p.

32. Simon, J. E., (1990). Essential oils and culinary herbs. In: *Advances in New Crops. Proceedings of the First National Symposium' New Crops: Research, Development, Economics,' Indianapolis, Indiana, USA, 1988.* (pp. 472–483). Timber Press.
33. Bakkali, F., Averbeck, S., Averbeck, D., & Idaomar, M., (2008). Biological effects of essential oils: a review. *Food and Chemical Toxicology, 46*(2), 446–475.
34. Burt, S., (2004). Essential oils: Their antibacterial properties and potential applications in foods—a review. *International Journal of Food Microbiology, 94*(3), 223–253.
35. Swamy, M. K., Mohanty, S. K., Sinniah, U. R., & Maniyam, A., (2015). Evaluation of patchouli (*Pogostemon cablin* Benth.) cultivars for growth, yield and quality parameters. *Journal of Essential Oil Bearing Plants, 18*(4), 826–832.
36. Tajkarimi, M. M., Ibrahim, S. A., & Cliver, D. O., (2010). Antimicrobial herb and spice compounds in food. *Food Control, 21*(9), 1199–1218.
37. Nazzaro, F., Fratianni, F., De Martino, L., Coppola, R., & De Feo, V., (2013). Effect of essential oils on pathogenic bacteria. *Pharmaceuticals, 6*(12), 1451–1474.
38. Lang, G., & Buchbauer, G., (2012). A review on recent research results (2008–2010) on essential oils as antimicrobials and antifungals. A review. *Flavor and Fragrance Journal, 27*(1), 13–39.
39. Sell, C., (2006). *The Chemistry of Fragrances: From Perfumer to Consumer* (Vol. 38). Royal Society of Chemistry.
40. Böhme, K., Barros-Velázquez, J., Calo-Mata, P., & Aubourg, S. P., (2014). Antibacterial, antiviral and antifungal activity of essential oils: Mechanisms and applications. In: *Antimicrobial Compounds* (pp. 51–81). Springer, Berlin Heidelberg.
41. Swamy, M. K., Akhtar, M. S., & Sinniah, U. R., (2016). Antimicrobial properties of plant essential oils against human pathogens and their mode of action: An updated review. *Evidence-Based Complementary and Alternative Medicine*.
42. Dorman, H. J. D., & Deans, S. G., (2000). Antimicrobial agents from plants: Antibacterial activity of plant volatile oils. *Journal of Applied Microbiology, 88*(2), 308–316.
43. Ultee, A., Bennik, M. H. J., & Moezelaar, R., (2002). The phenolic hydroxyl group of carvacrol is essential for action against the food-borne pathogen Bacillus cereus. *Applied and Environmental Microbiology, 68*(4), 1561–1568.
44. Ben Arfa, A., Combes, S., Preziosi☐Belloy, L., Gontard, N., & Chalier, P., (2006). Antimicrobial activity of carvacrol related to its chemical structure. *Letters in Applied Microbiology, 43*(2), 149–154.
45. Laekeman, G. M., Van Hoof, L., Haemers, A., Berghe, D. A., Herman, A. G., & Vlietinck, A. J., (1990). Eugenol a valuable compound for *in vitro* experimental research and worthwhile for further *in vivo* investigation. *Phytotherapy Research, 4*(3), 90–96.
46. Zemek, J., Košíková, B., Augustin, J., & Joniak, D., (1979). Antibiotic properties of lignin components. *Folia Microbiologica, 24*(6), 483–486.
47. Zemek, J., Valent, M., Pódová, M., Košíková, B., & Joniak, D., (1987). Antimicrobial properties of aromatic compounds of plant origin. *Folia Microbiologica, 32*(5), 421–425.
48. Gill, A. O., & Holley, R. A., (2004). Mechanisms of bactericidal action of cinnamaldehyde against Listeria monocytogenes and of eugenol against *L. monocytogenes* and Lactobacillus sakei. *Applied and Environmental Microbiology, 70*(10), 5750–5755.

49. Thorosk, I. J., Blank, G., & Biliaderis, C., (1989). Eugenol induced inhibition of extracellular enzyme production by Bacillus subtilis. *Journal of Food Protection, 52*(6), 399–403.

50. Mrozik, A., Piotrowska-Seget, Z., & Łabużek, S., (2004). Changes in whole cell-derived fatty acids induced by naphthalene in bacteria from genus Pseudomonas. *Microbiological Research, 159*(1), 87–95.

51. Gustafson, J. E., Liew, Y. C., Chew, S., Markham, J., Bell, H. C., Wyllie, S. G., et al., (1998). Effects of tea tree oil on Escherichia coli. *Letters in Applied Microbiology, 26*(3), 194–198.

52. Juven, B. J., Kanner, J., Schved, F., & Weisslowicz, H., (1994). Factors that interact with the antibacterial action of thyme essential oil and its active constituents. *Journal of Applied Microbiology, 76*(6), 626–631.

53. Lambert, R. J. W., Skandamis, P. N., Coote, P. J., & Nychas, G. J., (2001). A study of the minimum inhibitory concentration and mode of action of oregano essential oil, thymol and carvacrol. *Journal of Applied Microbiology, 91*(3), 453–462.

54. Helander, I. M., Alakomi, H. L., Latva-Kala, K., Mattila-Sandholm, T., Pol, I., Smid, E. J., et al., (1998). Characterization of the action of selected essential oil components on Gram-negative bacteria. *Journal of Agricultural and Food Chemistry, 46*(9), 3590–3595.

55. Turina, A. D. V., Nolan, M. V., Zygadlo, J. A., & Perillo, M. A., (2006). Natural terpenes: Self-assembly and membrane partitioning. *Biophysical Chemistry, 122*(2), 101–113.

56. Tassou, C., Koutsoumanis, K., & Nychas, G. J., (2000). Inhibition of Salmonella enteritidis and Staphylococcus aureus in nutrient broth by mint essential oil. *Food Research International, 33*(3), 273–280.

57. Carson, C. F., Mee, B. J., & Riley, T. V., (2002). Mechanism of action of melaleuca alternifolia (tea tree) oil on staphylococcus aureus determined by time-kill, lysis, leakage, and salt tolerance assays and electron microscopy. *Antimicrobial Agents and Chemotherapy, 46*(6), 1914–1920.

58. Turker, H., & Yıldırım, A. B., (2015). Screening for antibacterial activity of some Turkish plants against fish pathogens: A possible alternative in the treatment of bacterial infections. *Biotechnology & Biotechnological Equipment, 29*(2), 281–288.

59. Khan, R. A., Khan, F., Ahmed, M., Shah, A. S., Khan, N. A., Khan, M. R., et al., (2011). Phytotoxic and antibacterial assays of crude methanolic extract of *Mentha longifolia* (Linn.). *African Journal of Pharmacy and Pharmacology, 5*(12), 1530–1533.

60. Demirci, F., Guven, K., Demirci, B., Dadandi, M. Y., & Baser, K. H. C., (2008). Antibacterial activity of two Phlomis essential oils against food pathogens. *Food Control, 19*(12), 1159–1164.

61. Pelczar, M. J., Chan, E. C. S., & Krieg, N. R., (1993). *Microbiology: Concepts and Applications*, 80–100.

62. Pauli, A., (2002). Antimicrobial properties of catechol derivatives. In: *3rd World Congress on Allelopathy* (pp. 26–30). Tsukuba, Japan.

63. Noguchi, T., Hwang, D. F., Arakawa, O., Sugita, H., Deguchi, Y., Shida, Y., & Hashimoto, K., (1987). Vibrio alginolyticus, a tetrodotoxin-producing bacterium, in the intestines of the fish Fugu vermicularis vermicularis. *Marine Biology, 94*(4), 625–630.
64. Thomas, J., Jerobin, J., Seelan, T. S. J., Thanigaivel, S., Vijayakumar, S., Mukherjee, A., et al., (2013). Studies on pathogenecity of *Aeromonas salmonicida* in catfish *Clarias batrachus* and control measures by neem nanoemulsion. *Aquaculture, 396*, 71–75.
65. Mishra, P., Jerobin, J., Thomas, J., Mukherjee, A., & Chandrasekaran, N., (2014). Study on antimicrobial potential of neem oil nanoemulsion against Pseudomonas aeruginosa infection in Labeo rohita. *Biotechnology and Applied Biochemistry, 61*(5), 611–619.
66. Thomas, J., Thanigaivel, S., Vijayakumar, S., Acharya, K., Shinge, D., Seelan, T. S. J., et al., (2014). Pathogenecity of pseudomonas aeruginosa in *Oreochromis mossambicus* and treatment using lime oil nanoemulsion. *Colloids and Surfaces B: Biointerfaces, 116*, 372–377.
67. Ghosh, V., Mukherjee, A., & Chandrasekaran, N., (2013). Ultrasonic emulsification of food-grade nanoemulsion formulation and evaluation of its bactericidal activity. *Ultrasonics Sonochemistry, 20*(1), 338–344.
68. Nirmala, M. J., Allanki, S., Mukherjee, A., & Chandrasekaran, N., (2013). Azithromycin: Essential oil based nanoemulsion drug delivery system. *Int. J. Pharm. Pharm. Sci., 5*, 273–275.
69. Jo, Y. J., Chun, J. Y., Kwon, Y. J., Min, S. G., Hong, G. P., & Choi, M. J., (2015). Physical and antimicrobial properties of trans-cinnamaldehyde nanoemulsions in watermelon juice. *LWT-Food Science and Technology, 60*(1), 444–451.
70. Ghaderi-Ghahfarokhi, M., Barzegar, M., Sahari, M. A., Gavlighi, H. A., & Gardini, F., (2017). Chitosan-cinnamon essential oil nano-formulation: Application as a novel additive for controlled release and shelf life extension of beef patties. *International Journal of Biological Macromolecules, 102*, 19–28.
71. Sugumar S, Clarke S K, Nirmala M K, Tyagi B K, Mukherjee A, Chandrasekaran N et a.l (2013) Nanoemulsion of eucalyptus oil and its larvicidal activity against *Culex quinquefasciatus, Bull Entomol Res* 104, 393–402.
72. Alam, M. S., Ali, M. S., Alam, N., Siddiqui, M. R., Shamim, M., & Safhi, M. M., (2013). In vivo study of clobetasol propionate loaded nanoemulsion for topical application in psoriasis and atopic dermatitis. *Drug Invention Today, 5*(1), 8–12.
73. Donsì, F., Annunziata, M., Vincensi, M., & Ferrari, G., (2012). Design of nanoemulsion-based delivery systems of natural antimicrobials: Effect of the emulsifier. *Journal of Biotechnology, 159*(4), 342–350.
74. Donsì, F., Annunziata, M., Sessa, M., & Ferrari, G., (2011). Nanoencapsulation of essential oils to enhance their antimicrobial activity in foods. *LWT-Food Science and Technology, 44*(9), 1908–1914.
75. Zahi, M. R., Wan, P., Liang, H., & Yuan, Q., (2014). Formation and stability of D-limonene organogel-based nanoemulsion prepared by a high-pressure homogenizer. *Journal of Agricultural and Food Chemistry, 62*(52), 12563–12569.
76. Zhang, Z., Vriesekoop, F., Yuan, Q., & Liang, H., (2014). Effects of nisin on the antimicrobial activity of D-limonene and its nanoemulsion. *Food Chemistry, 150*, 307–312.

77. Hilbig, J., Ma, Q., Davidson, P. M., Weiss, J., & Zhong, Q., (2016). Physical and antimicrobial properties of cinnamon bark oil co-nanoemulsified by lauric arginate and Tween 80. *International Journal of Food Microbiology, 233*, 52–59.

78. Anwer, M. K., Jamil, S., Ibnouf, E. O., & Shakeel, F., (2014). Enhanced antibacterial effects of clove essential oil by nanoemulsion. *Journal of Oleo Science, 63*(4), 347–354.

79. Shahavi, M. H., Hosseini, M., Jahanshahi, M., Meyer, R. L., & Darzi, G. N., (2016). Clove oil nanoemulsion as an effective antibacterial agent: Taguchi optimization method. *Desalination and Water Treatment, 57*(39), 18379–18390.

80. Majeed, H., Antoniou, J., Shoemaker, C. F., & Fang, Z., (2015). Action mechanism of small and large molecule surfactant-based clove oil nanoemulsions against foodborne pathogens and real-time detection of their subpopulations. *Archives of Microbiology, 197*(1), 35–45.

81. Luo, Y., Zhang, Y., Pan, K., Critzer, F., Davidson, P. M., & Zhong, Q., (2014). Self-emulsification of alkaline-dissolved clove bud oil by whey protein, gum arabic, lecithin, and their combinations. *Journal of Agricultural and Food Chemistry, 62*(19), 4417–4424.

82. Terjung, N., Löffler, M., Gibis, M., Hinrichs, J., & Weiss, J., (2012). Influence of droplet size on the efficacy of oil-in-water emulsions loaded with phenolic antimicrobials. *Food & Function, 3*(3), 290–301.

83. Nuchuchua, O., Sakulku, U., Uawongyart, N., Puttipipatkhachorn, S., Soottitantawat, A., & Ruktanonchai, U. (2009). In vitro characterization and mosquito (Aedes aegypti) repellent activity of essential-oils-loaded nanoemulsions. *Aaps Pharmscitech, 10*(4), 1234.

84. Sakulku, U., Nuchuchua, O., Uawongyart, N., Puttipipatkhachorn, S., Soottitantawat, A., & Ruktanonchai, U., (2009). Characterization and mosquito repellent activity of citronella oil nanoemulsion. *International Journal of Pharmaceutics, 372*(1), 105–111.

85. Lu, W. C., Huang, D. W., Wang, C. C., Yeh, C. H., Tsai, J. C., Huang, Y. T., et al., (2017). Preparation, characterization, and antimicrobial activity of nanoemulsions incorporating citral essential oil. *Journal of Food and Drug Analysis, 26*(1), 82-89.

86. Salvia-Trujillo, L., Rojas-Graü, M. A., Soliva-Fortuny, R., & Martín-Belloso, O., (2014a). Formulation of antimicrobial edible nanoemulsions with pseudo-ternary phase experimental design. *Food and Bioprocess Technology, 7*(10), 3022–3032.

87. Salvia-Trujillo, L., Rojas-Graü, M. A., Soliva-Fortuny, R., & Martín-Belloso, O., (2014b). Impact of microfluidization or ultrasound processing on the antimicrobial activity against Escherichia coli of lemongrass oil-loaded nanoemulsions. *Food Control, 37*, 292–297.

88. Bonferoni, M. C., Sandri, G., Rossi, S., Usai, D., Liakos, I., Garzoni, A., et al., (2017). A novel ionic amphiphilic chitosan derivative as a stabilizer of nanoemulsions: Improvement of antimicrobial activity of Cymbopogon citratus essential oil. *Colloids and Surfaces B: Biointerfaces, 152*, 385–392.

89. Adhavan, P., Kaur, G., Princy, A., & Murugan, R., (2017). Essential oil nanoemulsions of wild patchouli attenuate multi-drug resistant gram-positive, gram-negative and Candida albicans. *Industrial Crops and Products, 100*, 106–116.

90. Jufri, M., & Natalia, M., (2014). Physical stability and antibacterial activity of black cumin oil (Nigella sativa L.) nanoemulsion Gel. *International Journal of Pharm. Tech. Research, 6*(4), 1162–1169.

91. Sharif, H. R., Abbas, S., Majeed, H., Safdar, W., Shamoon, M., Khan, M. A., et al., (2017). Formulation, characterization and antimicrobial properties of black cumin essential oil nanoemulsions stabilized by OSA starch. *Journal of Food Science and Technology, 54*(10), 3358–3365.

92. Shaaban, H. A., Sadek, Z., Edris, A. E., & Saad-Hussein, A., (2015). Analysis and antibacterial activity of Nigella sativa essential oil formulated in microemulsion system. *Journal of Oleo Science, 64*(2), 223–232.

93. Salvia-Trujillo, L., Rojas-Graü, A., Soliva-Fortuny, R., & Martín-Belloso, O., (2015). Physicochemical characterization and antimicrobial activity of food-grade emulsions and nanoemulsions incorporating essential oils. *Food Hydrocolloids, 43*, 547–556.

94. Moghimi, R., Ghaderi, L., Rafati, H., Aliahmadi, A., & McClements, D. J., (2016). Superior antibacterial activity of nanoemulsion of Thymus daenensis essential oil against E. coli. *Food Chemistry, 194*, 410–415.

95. Wu, J. E., Lin, J., & Zhong, Q., (2014). Physical and antimicrobial characteristics of thyme oil emulsified with soluble soybean polysaccharide. *Food Hydrocolloids, 39*, 144–150.

95. Xue, J., Davidson, P. M., & Zhong, Q., (2015). Antimicrobial activity of thyme oil co-nanoemulsified with sodium caseinate and lecithin. *International Journal of Food Microbiology, 210*, 1–8.

96. Bajerski, L., Michels, L. R., Colomé, L. M., Bender, E. A., Freddo, R. J., Bruxel, F., et al., (2016). The use of Brazilian vegetable oils in nanoemulsions: An update on preparation and biological applications. *Brazilian Journal of Pharmaceutical Sciences, 52*(3), 347–363.

97. Constantinides, P. P., Chaubal, M. V., & Shorr, R., (2008). Advances in lipid nanodispersions for parenteral drug delivery and targeting. *Advanced Drug Delivery Reviews, 60*(6), 757–767.

98. Wang, L., Tabor, R., Eastoe, J., Li, X., Heenan, R. K., & Dong, J., (2009). Formation and stability of nanoemulsions with mixed ionic–nonionic surfactants. *Physical Chemistry Chemical Physics, 11*(42), 9772–9778.

99. Calderó, G., García-Celma, M. J., & Solans, C., (2011). Formation of polymeric nanoemulsions by a low-energy method and their use for nanoparticle preparation. *Journal of Colloid and Interface Science, 353*(2), 406–411.

100. Gupta, A., Eral, H. B., Hatton, T. A., & Doyle, P. S., (2016). Nanoemulsions: Formation, properties and applications. *Soft Matter, 12*(11), 2826–2841.

101. Hörmann, K., & Zimmer, A., (2016). Drug delivery and drug targeting with parenteral lipid nanoemulsions—A review. *Journal of Controlled Release, 223*, 85–98.

102. Tadros, T., Izquierdo, P., Esquena, J., & Solans, C., (2004). Formation and stability of nano-emulsions. *Advances in Colloid and Interface Science, 108*, 303–318.

103. Myers, D., (1999). Wetting and spreading. *Surfaces, Interfaces, and Colloids: Principles and Applications, Second Edition*, 415–447.

104. Mason, T. G., Wilking, J. N., Meleson, K., Chang, C. B., & Graves, S. M., (2006). Nanoemulsions: Formation, structure, and physical properties. *Journal of Physics: Condensed Matter, 18*(41), R635.

105. Date, A. A., & Nagarsenker, M. S., (2008). Parenteral microemulsions: an overview. *International Journal of Pharmaceutics, 355*(1), 19–30.

106. Izquierdo, P., (2002). *Studies on Nano-Emulsion Formation and Stability* (Doctoral dissertation, Thesis, University of Barcelona, Spain).

107. Nam, Y. S., Kim, J. W., Shim, J., Han, S. H., & Kim, H. K., (2010). Nanosized emulsions stabilized by semisolid polymer interphase. *Langmuir, 26*(16), 13038–13043.

108. Wooster, T. J., Golding, M., & Sanguansri, P., (2008). Impact of oil type on nanoemulsion formation and Ostwald ripening stability. *Langmuir, 24*(22), 12758–12765.

109. Nemen, D., & Lemos-Senna, E., (2011). Preparation and characterization of resveratrol-loaded lipid-based nanocarriers for cutaneous administration. Quim. *Química Nova., 34*(3), 408–413.

110. Anton, N., Benoit, J. P., & Saulnier, P., (2008). Design and production of nanoparticles formulated from nano-emulsion templates—a review. *Journal of Controlled Release, 128*(3), 185–199.

111. Yukuyama, M. N., Ghisleni, D. D. M., Pinto, T. J. A., & Bou☐Chacra, N. A., (2016). Nanoemulsion: Process selection and application in cosmetics-a review. *International Journal of Cosmetic Science, 38*(1), 13–24.

112. Maruno, M., & Rocha-Filho, P. A. D., (2009). O/W nanoemulsion after 15 years of preparation: A suitable vehicle for pharmaceutical and cosmetic applications. *Journal of Dispersion Science and Technology, 31*(1), 17–22.

113. Taylor, P., (2003). Ostwald ripening in emulsions: Estimation of solution thermodynamics of the disperse phase. *Advances in Colloid and Interface Science, 106*(1), 261–285.

114. Urbina-Villalba, G., Forgiarini, A., Rahn, K., & Lozsán, A., (2009). Influence of flocculation and coalescence on the evolution of the average radius of an O/W emulsion. Is a linear slope of R [combining macron] 3 vs. t an unmistakable signature of Ostwald ripening? *Physical Chemistry Chemical Physics, 11*(47), 11184–11195.

115. Lifshitz, I. M., & Slyozov, V. V., (1961). The kinetics of precipitation from supersaturated solid solutions. *Journal of Physics and Chemistry of Solids, 19*(1–2), 35–50.

116. Pey, C. M., Maestro, A., Solé, I., González, C., Solans, C., & Gutiérrez, J. M., (2006). Optimization of nano-emulsions prepared by low-energy emulsification methods at constant temperature using a factorial design study. *Colloids and Surfaces A: Physicochemical and Engineering Aspects, 288*(1), 144–150.

117. Solè, I., Pey, C. M., Maestro, A., González, C., Porras, M., Solans, C., et al., (2010). Nano-emulsions prepared by the phase inversion composition method: Preparation variables and scale up. *Journal of Colloid and Interface Science, 344*(2), 417–423.

118. Karthik, P., Ezhilarasi, P. N., & Anandharamakrishnan, C., (2017). Challenges associated in stability of food grade nanoemulsions. *Critical Reviews in Food Science and Nutrition, 57*(7), 1435–1450.

119. Kourniatis, L. R., Spinelli, L. S., Mansur, C. R., & González, G., (2010). Nanoemulsões óleo de laranja/água preparadas em homogenizador de alta pressão. *Quim. Nova, 33*(2), 295–300.

120. Puglia, C., Rizza, L., Drechsler, M., & Bonina, F., (2010). Nanoemulsions as vehicles for topical administration of glycyrrhetic acid: Characterization and *in vitro* and *in vivo* evaluation. *Drug Delivery*, *17*(3), 123–129.

121. Sharma, N., Bansal, M., Visht, S., Sharma, P. K., & Kulkarni, G. T., (2010). Nanoemulsion: A new concept of delivery system. *Chronicles of Young Scientists*, *1*(2), 2.

122. Mahdi, J. S., He, Y., & Bhandari, B., (2006). Nano-emulsion production by sonication and microfluidization-a comparison. *International Journal of Food Properties*, *9*(3), 475–485.

123. Li, M. K., & Fogler, H. S., (1978). Acoustic emulsification. Part 2. Breakup of the large primary oil droplets in a water medium. *Journal of Fluid Mechanics*, *88*(3), 513–528.

124. Kentish, S., Wooster, T. J., Ashokkumar, M., Balachandran, S., Mawson, R., & Simons, L., (2008). The use of ultrasonics for nanoemulsion preparation. *Innovative Food Science & Emerging Technologies*, *9*(2), 170–175.

125. McClements, D. J., (2011). Edible nanoemulsions: Fabrication, properties, and functional performance. *Soft Matter*, *7*(6), 2297–2316.

126. Olson, D. W., White, C. H., & Richter, R. L., (2004). Effect of pressure and fat content on particle sizes in microfluidized milk. *Journal of Dairy Science*, *87*(10), 3217–3223.

127. Dalgleish, D. G., Tosh, S. M., & West, S., (1996). Beyond homogenization: The formation of very small emulsion droplets during the processing of milk by a microfluidizer. *Nederlands Melk en Zuiveltijdschrift*, *50*(2), 135–148.

128. Fortunato, E., (2005). As metas da nanotecnologia: Aplicações e Implicações. *Lisboa: Universidade Nova de Lisboa.*

129. Bilbao-Sáinz, C., Avena-Bustillos, R. J., Wood, D. F., Williams, T. G., & McHugh, T. H., (2010).Nanoemulsions prepared by a low-energy emulsification method applied to edible films. *Journal of Agricultural and Food Chemistry*, *58*(22), 11932–11938.

130. Solè, I., Maestro, A., Pey, C. M., González, C., Solans, C., & Gutiérrez, J. M. (2006). Nano-emulsions preparation by low energy methods in an ionic surfactant system. *Colloids and Surfaces A: Physicochemical and Engineering Aspects*, *288*(1), 138–143.

131. Sajjadi, S., (2006). Nanoemulsion formation by phase inversion emulsification: On the nature of inversion. *Langmuir*, *22*(13), 5597–5603.

132. McClements, D. J., (2012). Crystals and crystallization in oil-in-water emulsions: Implications for emulsion-based delivery systems. *Advances in Colloid and Interface Science*, *174*, 1–30.

133. Pouton, C. W., & Porter, C. J., (2008). Formulation of lipid-based delivery systems for oral administration: Materials, methods and strategies. *Advanced Drug Delivery Reviews*, *60*(6), 625–637.

134. Shinoda, K., & Saito, H., (1969). The stability of O/W type emulsions as functions of temperature and the HLB of emulsifiers: the emulsification by PIT-method. *Journal of Colloid and Interface Science*, *30*(2), 258–263.

135. Forgiarini, A., Esquena, J., Gonzalez, C., & Solans, C., (2001). Formation of nanoemulsions by low-energy emulsification methods at constant temperature. *Langmuir*, *17*(7), 2076–2083.

136. Morales, D., Gutiérrez, J. M., Garcia-Celma, M. J., & Solans, Y. C., (2003). A study of the relation between bicontinuous microemulsions and oil/water nano-emulsion formation. *Langmuir, 19*(18), 7196–7200.

137. Lovelyn, C., & Attama, A. A., (2011). Current state of nanoemulsions in drug delivery. *Journal of Biomaterials and Nanobiotechnology, 2*(05), 626.

138. Narang, A. S., Delmarre, D., & Gao, D., (2007). Stable drug encapsulation in micelles and microemulsions. *International Journal of Pharmaceutics, 345*(1), 9–25.

139. Gursoy, R. N., & Benita, S., (2004). Self-emulsifying drug delivery systems (SEDDS) for improved oral delivery of lipophilic drugs. *Biomedicine & Pharmacotherapy, 58*(3), 173–182.

140. Gershanik, T., Benzeno, S., & Benita, S., (1998). Interaction of a self-emulsifying lipid drug delivery system with the everted rat intestinal mucosa as a function of droplet size and surface charge. *Pharmaceutical Research, 15*(6), 863–869.

141. Ghosh, P. K., Majithiya, R. J., Umrethia, M. L., & Murthy, R. S., (2006). Design and development of microemulsion drug delivery system of acyclovir for improvement of oral bioavailability. *AAPS Pharmscitech, 7*(3), 172–177.

142. Lawrence, M. J., & Rees, G. D., (2012). Microemulsion-based media as novel drug delivery systems. *Advanced Drug Delivery Reviews, 64*, 175–193.

143. Chiesa, M., Garg, J., Kang, Y. T., & Chen, G., (2008). Thermal conductivity and viscosity of water-in-oil nanoemulsions. *Colloids and Surfaces A: Physicochemical and Engineering Aspects, 326*(1), 67–72.

144. Craig, D. Q. M., Barker, S. A., Banning, D., & Booth, S. W., (1995). An investigation into the mechanisms of self-emulsification using particle size analysis and low frequency dielectric spectroscopy. *International Journal of Pharmaceutics, 114*(1), 103–110.

145. Debnath, S., Rayana, S., & Vijay Kumar, G., (2011). Nanoemulsion-a method to improve the solubility of lipophilic drugs. *Pharmanest, 2*(2–3), 72–83.

146. Samah, N. A., Williams, N., & Heard, C. M., (2010). Nanogel particulates located within diffusion cell receptor phases following topical application demonstrates uptake into and migration across skin. *International Journal of Pharmaceutics, 401*(1), 72–78.

147. Shakeel, F., Baboota, S., Ahuja, A., Ali, J., & Shafiq, S., (2008). Skin permeation mechanism of Aceclofenac using novel nanoemulsion formulation. *Die Pharmazie: An International Journal of Pharmaceutical Sciences, 63*(8), 580–584.

148. Baboota, S., Shakeel, F., Ahuja, A., Ali, J., & Shafiq, S., (2007). Design, development and evaluation of novel nanoemulsion formulations for transdermal potential of celecoxib. *Acta Pharmaceutica, 57*(3), 315–332.

# INDEX